American Farmland Preservation

This book examines modern American farmland preservation, arguing that its policies must address the social, identity, and psychological issues facing American farmers.

As American farmland is disappearing rapidly, this crisis not only jeopardizes American agriculture but also threatens rural communities. While the American farmland preservation movement has focused on conservation easements, permanent land restrictions that work within the American legal system's preference for inalienable property rights, the author argues for a new era of farmland preservation based on stakeholder-led solutions. The book showcases how rapid development is escalating farmland loss in urbanizing parts of the country, meaning the problem facing much of rural America is the loss of farmers and farm communities, not farmland. Drawing on the author's career as a farmland preservation specialist, the book evaluates America's current farmland preservation policies framed through the "politics of place," a term describing the role of place in local politics and civic engagement. Utilizing interviews with farmland preservationists, the book highlights a broader, more holistic approach to farmland preservation than mere land protection. By featuring these prominent examples from around the country, the book puts forward a new vision for farmland preservation that more insightfully addresses the social and economic issues facing America's farming communities and fosters a more engaged rural citizenship.

This book will be of great interest to students and scholars of food and agriculture, rural sociology, and environmental management, as well as professionals working in farmland management, conservation, and preservation.

Andrew Waters completed his Ph.D. in natural resources policy from Clemson University in 2022 after a twenty-year career in land conservation and land trust leadership. Today, Waters works full-time as the farmland preservation coordinator for Chatham County, a rapidly urbanizing area in North Carolina's dynamic "Triangle" region, and serves as lecturer in the Master of Public Administration program at Clemson University, where he has taught a self-designed course on sustainable land use and adaptive government.

Earthscan Food and Agriculture

Agricultural Land Use
Structural Transformations, Environment Challenges, Planning and Policy
Edited by Jerzy Bański

Sustainable Production of Root and Tuber Crops
Edited by Ravinder Kumar, Milan Kumar Lal, Rahul Kumar Tiwari and Brajesh Singh

Agrifood Regimes in Greece
Sociotechnical Dynamics and Sustainable Transitions
Stathis Arapostathis, Sotiris Alexakis, Vasiliki Karantzavelou and Yannis Fotopoulos

Nourishing Scotland
Pete Ritchie

Community Supported Agriculture in Central and Eastern Europe
Transitional Economies in Post-Communist Europe
Magdalena Raftowicz and Mirosław Struś

Decolonising Seed Governance in the Andes
Lessons for Just and Sustainable Food Futures
Juan Garzón

Nature-Based Solutions for European Vineyards
Solutions for Transformative Change
Edited by Udaya Sekhar Nagothu

American Farmland Preservation
Policies, Practices, and the Politics of Place
Andrew Waters

For more information about this series, please visit: www.routledge.com/Earthscan-Food-and-Agriculture/book-series/ECEFA

American Farmland Preservation

Policies, Practices, and the Politics of Place

Andrew Waters

Routledge
Taylor & Francis Group
LONDON AND NEW YORK

earthscan
from Routledge

Designed cover image: Getty Images

First published 2027
by Routledge
4 Park Square, Milton Park, Abingdon, Oxon OX14 4RN

and by Routledge
605 Third Avenue, New York, NY 10158

Routledge is an imprint of the Taylor & Francis Group, an informa business

British Library Cataloguing-in-Publication Data
A catalogue record for this book is available from the British Library

ISBN: 978-1-041-08256-9 (hbk)
ISBN: 978-1-041-08255-2 (pbk)
ISBN: 978-1-003-64454-5 (ebk)

DOI: 10.4324/9781003644545

Typeset in Optima
by Taylor & Francis Books

For my patient and supportive doctoral advisors, Dr. Charles Chancellor and Dr. Lori Dickes.

And for Anne and Eli, as always.

Contents

Acknowledgments

Many of the ideas and themes of this book originated during my doctoral program in the Clemson University Department of Parks, Recreation, and Tourism Management. For their patient support and encouragement, I would especially like to thank my doctoral advisors, Dr. Charles Chancellor and Dr. Lori Dickes.

I would also like to acknowledge and thank my colleagues in the Chatham Soil and Water Conservation District Susannah Goldston, Brandy Oldham, and Joe Trimnal. Thank you for reminding me about the importance of our work and our mission of helping Chatham County farmers every day.

One of the true pleasures in writing this book was meeting the many talented farmland preservation practitioners that allowed me to interview them for this book. Specifically, I would like to thank the following people and their organizations: Ryan Anderson, Steward; Rosie Pahl Donaldson and Remy Long, Ann Arbor (MI) Greenbelt Program; Ariel Zjip, Buncombe County (NC) Soil and Water Conservation District; Nancy Stangle, Joe Barbaree, and Nigel Pritchard, Athens Land Trust; Benton Montgomery, GrowFood Carolina; Kavita Koppa, Rural Agricultural Foundation International; Chris Boggs, Washington County (MD) Rural Preservation Administrator; Kristián Varsa, Farms Fund (The Conservation Fund); Dane White, California Agricultural Leadership Foundation; David Howard, National Young Farmers Coalition; Holly Rippon Butler; Gabrielle Roesch-McNally, Women for the Land (American Farmland Trust); Jennifer Dempsey, Farmland Information Center (American Farmland Trust); Stephanie Kreiser, State Agriculture Development Committee (New Jersey); David Miller and Dr. Stephen Rivard, Iroquois Valley Farmland REIT. Thank you for sharing your insight and inspiration.

1 Farmland Preservation

A Wicked Problem

Albert Einstein may not have been referring specifically to wicked problems when he said, "we cannot solve our problems with the same level of thinking that created them," but the sentiment is as true today as it was a hundred years ago. Today, we use the term "wicked problems" to describe societal challenges that are resistant to consensus agreement, clear solutions, and identifiable issues, such as climate change, overpopulation, and poverty.[1] And if farmland preservation is one of America's current wicked problems, we can attribute it to no single cause, except perhaps centuries of exploitation. Population growth, urban sprawl, and the industrialization of American agriculture are surely contributing factors, but so are globalizing markets, declining social capital, and media digitalization, along with a host of other social, cultural, and economic issues, some of them dealing directly with agriculture and farmland, and some of them endemic to broader issues in American society.

From this perspective, we can use farmland preservation as a boundary object for a host of issues influencing American agriculture, all contributing to the wicked problem of declining rural communities, and the range of environmental, social, economic, and cultural problems that decline creates. Here, the term "boundary object" describes a concept or term situated at the intersection of research, policy, and disciplinary settings.[2] True, for many, American farmland preservation refers specifically to conservation easements and other tools of the American conservation movement. This book certainly addresses these "traditional" tools of farmland preservation, yet it also attempts to broaden the concept of farmland preservation to new and emerging practices, such as "buy-sell-lease" and farmland "commons" models, as well as different approaches to American democracy itself.

I first started working in farmland preservation in 2004, when I became the land protection director for the Catawba Lands Conservancy in the rapidly growing metro area of Charlotte, North Carolina. Today I serve as the farmland preservation director for Chatham County, North Carolina, an urbanizing county in North Carolina's equally booming "Triangle" region. In between I've participated in the "preservation" of dozens of working landscapes—farmland and working forests, along with the occasional hobby farm—primarily through

DOI: 10.4324/9781003644545-1

permanent conservation easements. In some ways, this book is an effort to synthesize my practitioner perspective on farmland preservation, explain and elaborate on some of the things I've learned over the years, and humbly propose revisions to America's current farmland preservation policy that better address its underlying issues.

Much has changed in farmland preservation since 2004, some for the better, some for the worse. For one, there is more funding available for farmland preservation now than there was back then. Another is the emergence of state and local government programs supporting farmland preservation, such as the one I manage in Chatham County. Many of these programs work in conjunction with local land trusts to acquire permanent farmland conservation easements from willing landowners. That's great news, but I worry these programs are creating the perception of conservation easements as a "one-size-fits-all" solution for farmland preservation, not a single, albeit useful, tool for a problem that requires many different tools to fix.

For those of you who may not know what conservation easement are, put simply, they are permanent deed restrictions that limit development and subdivision on a specific piece of land to protect its conservation values, including its prime soils, fields and forests, and other agricultural resources. As a land protection specialist, I've learned a lot about conservation easements over the last two decades; enough to be cautious with their use, and enough to know there is much about them I still do not know. And my practitioner perspective on farmland preservation has been informed by my scholarly studies in land conservation policy and practice. It was as a doctoral student that I was first exposed to the sociological theory of place, or sense of place, which the early place theorist Y.F. Tuan defines broadly as the meanings and attachment to a setting held by an individual or group.[3] Part of the reason I found the place theories of Tuan and his predecessors so appealing was that they confirmed my own experiences working in land conservation. Conservation easements work best, I have found, when used to preserve deep cognitive and emotional connections between family and place that develop over generations of fellowship with the land. Narrative plays an important role in this process, and one of the things conservation easements do exceptionally well is preserve the narrative of family connection to land.

But if place theory satisfied my scholarly need to understand the cognitive processes fueling a family's decision to permanently conserve their farmland, the practitioner part of my brain wanted to understand how to operationalize that understanding into more effective farmland preservation practice. This led me to the concept of the "politics of place," and eventually to Daniel Kemmis's book titled *Community and the Politics of Place*.

For Kemmis, the politics of place is essentially the politics of civic virtue and participation. He argues that meaningful political decisions regarding community issues like farmland preservation must take place in a dialog of local citizens encompassing a diverse range of opinions and viewpoints. Only through the friction of this admittedly messy process can the trust necessary to craft and implement solutions to wicked problems develop, and only through the scale of local place is this type of relationship-building possible.

Community and the Politics of Place was written in 1990. Given the current state of American politics, and the divisiveness of our political system, I admit Kemmis's ideas can feel somewhat naive from a twenty-first-century perspective. Yet I can also argue that some degree of naivete is necessary for tackling wicked problems—the skeptics are good at complaining but rarely propose meaningful solutions—and that the seeds of Kemmis's recommendations have already taken root in the land conservation movement, which has successfully embraced the practice of democratic stakeholder process over the last thirty years.

Policy geeks like me like to talk about the pros and cons of stakeholder process and citizen participation, or moan about its disappearance from American civics. But for many, it is often difficult to understand what, exactly, the term "stakeholder process" means. Yet even if you aren't familiar with the concept of stakeholder process, there's a decent chance you've participated in it, perhaps in a public hearing or as part of a planning process for a new park or road or community center in your neighborhood. Essentially, stakeholder process is the idea that we create more effective, responsive, and adaptive governance solutions by engaging a variety of constituents and viewpoints in the decision-making process, whether that be in policy, planning, or management. In some ways, it is an effort to bring back the participatory democracy of earlier American generations.

Stakeholder groups are now ubiquitous in some parts of American government, especially in the governance of planning, where they are often formed to address land use issues ranging from gentrification of urban neighborhoods to recreational infrastructure designs to allocations of water rights, typically in the form of some sort of facilitated planning process. But Kemmis hits on a fundamental truth about stakeholder participation that a lot of these current efforts ignore—that it is only through the relatively flat scale of local politics that the principles of stakeholder participation can truly be effective in solving local problems, and then only through the sustained, regular interactions of day-to-day political process. "Places breed cooperation," Kemmis states. That is not to say that such cooperation comes easily, or is without conflict, but "any kind of active citizenship, where different interests are directly engaged in working out their problems and possibilities among themselves, is going to work better in small than in large political entities."[4]

As someone who has spent a fair amount of time knocking around farms and sitting in kitchens listening to stories about a family's connections to its land, as well as attending both local and statewide meetings about farmland preservation, Kemmis's observation rings true to me. Local communities need their own farmland preservation initiatives, tailored to their own needs and their own communities.

Why, then, does the current farmland preservation community insist on promoting the conservation easement as its "one-stop shop" for protecting farmland? And if I exaggerate here, for there are numerous efforts taking other, thoughtful approaches, just Google "farmland preservation" and see what appears on your computer screen—mostly, I'm betting, information about conservation easements or specific farms that were recently preserved through conservation easements. To some degree, that's because the use of conservation easements continues to explode across the United States, with 15 million acres protected through conservation easements in the years 2010 to 2020 alone.[5] And attendant to this rise in the popularity of conservation easements is a subsequent increase in government programs that provide financial assistance for conservation easement acquisitions. When targeted toward farmland preservation, these are known as PACE programs, an acronym for "Purchase of Agricultural Conservation Easements." Running a local PACE program for Chatham County is one of my primary responsibilities as its farmland preservation coordinator.

One reason for the popularity of conservation easements and PACE programs in America is that they fit nicely within our legal system of private property rights. We also have a long legal history of "severing" those rights, either for the economic interest of a third party or for some type of public use, such as a road or powerline rights of way. If you can sell and permanently convey the mineral or water or timber rights on your property, then why not the conservation rights? Assuming the entity you sell or convey them to has the ongoing capacity to protect those rights, or hold them in trust, this can ensure that the property will stay in its undeveloped state for perpetuity, or at least a very long time. Indeed, one of the reasons that conservation easements have become popular is that they adopt the same property rights arguments that many in America use to oppose widespread land-use regulations.

Give a politically-divided-yet-rapidly-growing county like Chatham a non-regulatory approach to meaningful land preservation, and it's no wonder they will embrace conservation easements and PACE programs. After all, residents here have been hearing about the success of conservation easements through the work of our outstanding local land trust for the last forty years. And they are also starting to hear about the Agricultural Development and Farmland Preservation Trust Fund, a statewide PACE program run through the North Carolina Department of Agriculture and Consumer Services that was established in the early 2000s.

Conservation easements are sexy. They are innovative. They use the same property rights laws and principles that contribute to the issues of unchecked development to create meaningful environmental protections. Furthermore, they provide politicians with hard numbers of acres protected and dollars invested in conservation and preservation. And they *can* achieve meaningful land-use protections without land-use regulations. It's no wonder governments like them, or at least find them politically expedient. But the truth is that a conservation easement is a very specific and very powerful tool that is only suitable for a small percentage of Americans. Yes, they do a good job protecting farmland, but they're not always great at meeting the changing needs of families and family businesses. And the danger of promoting them as *the* answer to our farmland preservation crisis is that they will lead to legal restrictions that may well protect the land itself but do little to address the underlying social and economic issues plaguing America's small and mid-size farming communities.

This book hypothesizes that adopting conservation easements as a type of national farmland preservation policy ignores the principles of the politics of place, the idea that politics works best on a local scale, and that meaningful political solutions must be crafted to fit the unique values and needs of that place, ideally through an engaged citizen process.

The purpose of this book, then, is to explore the factors leading to the crisis of farmland loss in America and gain a deeper understanding about the land use, legal, and sociological factors creating it. By no means is it intended as a "takedown" of conservation easement, but rather a call to action for a more adaptive and comprehensive approach to farmland preservation policy, one that addresses the place-based issues of identity, community, economics, and family dynamics. To achieve this, I first will explore farmland preservation as a natural resource, or land-use crisis, and investigate how a growing number of policies and organizations are attempting to address the farmland preservation crisis through a focus on the land itself. This will require a deeper look at the instruments and legal frameworks of land conservation, leading to an in-depth look at conservation easements, and programs that enable governments and their nonprofit partners to acquire conservation easements with public funds. We'll also explore and analyze other programs that attempt to "preserve" farmland itself, including agricultural district programs and farmland transition planning.

But then I move beyond the conventional boundaries of current farmland preservation practices to explore the broader sociological role of farmland in the American consciousness and the underlying social factors that are contributing to the farmland crisis. For this discussion, I'll use Kemmis's "politics of place" as a framework, focusing on how farmland preservation must adapt to serve the specific needs of the communities where these programs and policies operate. As we develop this analytical framework, we will go back and look at the origins of place as both a political and sociological construct, with a heavy dose of Montesquieu and Tocqueville as well. We'll also go forward, looking at how a generation of scholars and thinkers have adapted Kemmis's concepts for models of governance designed to address the profound change facing America's farming communities, as well as other aspects of American society. To do this, we'll explore more recent theoretical thought relating to place, social capital, and stakeholder governance.

Part One is titled "Foundations of American Farmland Preservation" and describes the history, legal issues, and theoretical framing of farmland preservation today. This exploration begins with an overview of the farmland loss crisis in Chapter 2. Chapter 3 provides a brief history of farmland preservation and agricultural land-use policy in America to the end of the twentieth century. Chapter 4 introduces the legal framework of American property rights that has brought us to the widespread use of conservation easements in farmland preservation policy, while Chapter 5 provides a similar introduction to the different types of American "agrarianism" and how their competing goals and objects influence American farmland preservation practice.

Part Two of the book is titled "American Farmland Preservation Practices and Policies Today" and explores the current state of the American farmland preservation movement. Chapter 6 looks broadly at current American farmland preservation practices, focusing on the role of conservation easements and the growing role of government acquisition easement programs. Chapter 7 takes a deeper dive into conservation easements: what they are, how they work, how they're funded, and why an American farming family may or may not want one. Chapter 8 explores how some local American places are attempting to adapt the traditional land-use policy tools of zoning and planning to the farmland preservation issue, with mixed success, while Chapter 9 looks at the issue from a statewide level, profiling specifically the farmland preservation programs of New Jersey and Maryland. Chapter 10 describes the issue of farmland transition, which refers to the planning process that is necessary to protect farmland in the dangerous ownership shift from one farming generation to the next. If conservation easements are the "superstars" of farmland preservation policy, farmland transition planning is its heart and soul, both more practical and essential than conservation easements, but also more existential because it forces families to face the complex issues surrounding mortality, fairness, and the fragility of human relationships.

Part Three is called "American Farmland Preservation and the Politics of Place," and it is here where I transition from the land-focused approaches of current American farmland preservation policy to explorations of social and place-based approaches. In Chapter 11, I briefly discuss social capital theory, which has evolved from the resource-based theories of French sociologists Pierre Bourdieu to the civic approach popularized by the latter-day sociologist Robert Putnam in his famous work *Bowling Alone*. For me, any discussion of social capital must accompany one of place theory. If our sense of place is typically characterized by our emotional and cognitive connections to the biophysical world—the physical features of environment, climate, landscape, etc.—social dimensions of place such as family, friendships, and governance also play a prominent role in influencing this relationship. Place theory is a broad, rich, and diverse field with research enough to fill many books. For my overview, I'll look briefly at the evolution of place theory, along with some related place concepts, before delving into the role I believe place plays in farmland preservation.

After exploring the theoretical constructs of social capital and sense of place introduced in Chapter 10, Chapter 12 introduces the "politics of place" concept, derived from the work of Daniel Kemmis in his influential 1990 book *Community and the Politics of Place*. But first I start with political theories of the French sociologist Montesquieu, who developed ideas about the role of place in politics back in the eighteenth century, and trace Montesquieu's influence through his disciple Alexis de Tocqueville's observations on the role of place in American politics to the place-based political models of Kemmis and other, latter-day place scholars. Chapter 13 explores new tools of farmland preservation, or new iterations of old tools, such as "Buy-Protect-Sell" and "Commons" models that, I believe, are more aligned with community-based farmland preservation, placing them within the politics of place paradigm.

Chapter 14 takes a look at the farmland preservation role of bridging organizations, which link people and ideas across hierarchies of government and are necessary for generating the social capital that can engender adaptive policy approaches. In farmland preservation, this bridging role is frequently played by land trusts, who are able to use their organizational resources to connect farmers and agricultural landowners throughout the broader conservation and government agency community—an important role in the politics of place. Chapter 15 explores the similar function of narrative networks, which use stories to connect people, ideas, and actions across social and institutional contexts. In farmland preservation, narrative networks employ stories of the challenges surrounding land access and land tenure to provide training, advocacy, and resources for those disproportionately afflicted by those challenges. Chapter 16 takes a brief look at a very specific form of narrative network, the Internet-based investment platforms that are using stories about farmland preservation and sustainable agriculture to fund the innovative small- and medium-scaled agricultural businesses that often constitute the heart of local agriculture revitalization efforts. A selective overview of farmland preservation policies and practices from around the globe is the topic of Chapter 17, followed by a brief summary and look ahead to the future of American farmland preservation in Chapter 18.

Admittedly, there is no silver bullet for American farmland preservation policy, just as there is no one answer for other wicked problems facing our country and the world. But it is not in our human nature to surrender to this wickedness. And solving wicked problems demands wicked solutions—transdisciplinary, multi-stakeholder, and adaptive responses that forego positivistic objectives and single-pronged solutions. If the relative success of the farmland preservation movement over the last thirty to forty years has focused on land conservation and resource protection as its primary tool, we are now ready to use that experience as the cornerstone for a more adaptive version of farmland protection that also addresses its economic, social, and cognitive dimensions. Even if we are just at the beginning of this journey, the time for farmland preservation centered in the politics of place is now.

Notes

1 Ruth DeFries and Harini Nagendra, "Ecosystem Management as a Wicked Problem," *Science* 356, no. 6335 (2017): 265–270.
2 J.A. Baggio, K. Brown, and D. Hellebrandt, "Boundary Object or Bridging Concept," *Ecology & Society*, 20, no. 2 (2015).
3 Y.F. Tuan, *Space and Place: The Perspective of Experience* (University of Minnesota Press, 1977).
4 Daniel Kemmis, *Community and the Politics of Place* (University of Oklahoma Press, 1990), 122, 125.
5 "61 Million Acres Voluntarily Conserved in America," Land Trust Alliance, December 2, 2021, https://landtrustalliance.org/newsroom/press-releases/61-million-acres-voluntarily-conserved-in-america-2020-national-land-trust-census-report-reveals.

Bibliography

Baggio, J.A., K. Brown, and D. Hellebrandt. 2015. "Boundary Object or Bridging Concept: A Citation Network Analysis of Resilience." *Ecology & Society*, 20, no. 2.

Defries, Ruth, and Harini Nagendra. 2017. "Ecosystem Management as a Wicked Problem." *Science* 356 (6335): 265–270. doi:10.1126/science.aal19.

Kemmis, Daniel. 1990. *Community and the Politics of Place*. University of Oklahoma Press.

Tuan, Y.F. 1977. *Space and Place: The Perspective of Experience*. University of Minnesota Press.

Part One

Foundations of American Farmland Preservation

2 A Tale of Two Farmlands

My community is losing farmland. And it's not just here, in Chatham County, North Carolina, where I work as the county's farmland preservation coordinator, but all over America. A recent study sponsored by the American Farmland Trust (AFT), a national nonprofit that works to preserve America's farmland resources, estimates that 18.4 million acres of farmland will be lost in the United States by 2040. And that's just the loss if we keep developing our farmland as we've done for the last few decades. Under AFT's "Runaway Sprawl" scenario, projected farmland loss by 2040 increases to 24.4 million acres.[1]

Of course, this loss is more drastic in states experiencing intense development pressure and population growth, and for better or worse, North Carolina is one of them. Under AFT's "Business as Usual" scenario, North Carolina is projected to lose 1,197,300 acres of farmland by 2040, with an additional loss of 480,800 acres under their "Runaway Sprawl" projections. This gives North Carolina the dubious distinction of ranking second on the AFT's "Top 12 Most Threatened States" list, behind only Texas and ahead of other urbanizing states like New Jersey, Florida, and Pennsylvania.[2]

For Chatham County—a part of North Carolina's rapidly urbanizing "Triangle" region that includes Raleigh, Durham, and Chapel Hill—the news doesn't get any better from there. AFT's analysis indicates that Chatham County will lose 19,400 acres of farmland by 2040 under its "Runaway Sprawl" scenario, approximately one-fifth of its current farmland. Much of this loss will be due to population growth. Between 2010 and 2022, Chatham County's population growth was 25.8 percent, compared to 11.7 percent for the entire state, and this population explosion is not projected to slow down any time soon. The North Carolina Office of State Budget and Management estimates that Chatham County's population will increase 17.8 percent by 2030, in line with or ahead of other rapidly urbanizing North Carolina counties like Durham (12 percent), Mecklenburg, the home of Charlotte (17.2 percent), and Buncombe, the home of Asheville (10 percent).[3]

This population growth is fueling intense residential development here in Chatham County. The 7,100-acre "Chatham Park" development in the county seat of Pittsboro is slated to be Chatham County's largest-ever

DOI: 10.4324/9781003644545-3

residential project, with up to 22,000 planned residential units. In 2024, the Walt Disney Company announced that Chatham Park would become the home of a Disney "Storyliving by Disney" branded residential community, with homes slated to go on the market by 2027, escalating fears about runaway residential development (and casting Mickey Mouse as an unlikely villain in the local farmland preservation movement).[4] Even as development increases, housing prices skyrocket. A recent newspaper article reports that median home prices in Chatham County have surpassed $600,000, compared to North Carolina's median sales price of $367,000.[5]

But the threat isn't only from residential development. Chatham County is also home to two of North Carolina's largest economic development projects: a 450-acre site for the semiconductor company Wolfspeed, billed as the world's largest silicon carbide manufacturing center; and the 2,000-acre Vinfast electric automobile manufacturing plant (the Vinfast plant was on hold as of the book's publication due to the financial issues and a softening in the electric car market, although company officials swear the project is still on track for later this decade).[6]

The dual pressure of residential and industrial development is driving Chatham County real estate prices through the roof. Chatham County's average listing price per undeveloped acre is $45,532, according to a local real estate website.[7] Although this average includes lots already subdivided for residential development, undeveloped farmland in Chatham County typically sells in the range of $15,000–$30,000 per acre, depending on location, a price far higher than most local farmers can afford to pay for it.

But in many ways, Chatham is a tale of two counties. The higher end of that price range is almost always found in the eastern half of the county, the part that abuts the metropolitan areas of Raleigh, Durham, and Chapel Hill. The county's eastern half is one of the few Triangle locations located within a convenient thirty-minute drive (on a good traffic day) to all three of these cities, making it an attractive location for commuters.

With some prominent exceptions, the problem facing the western half of Chatham County is not the loss of farmland but a lack of farmers to tend the vacant fields and pastures one encounters on its lonely country roads. If housing developments are growing like weeds in eastern Chatham County, the weeds are literal in parts of western Chatham County, which is struggling to maintain its agricultural economy.

In this dichotomy, Chatham County can serve as a microcosm of farming throughout America. If it is true that farmland is disappearing rapidly in urbanizing areas like the Triangle, and urbanizing states like North Carolina and Texas, it is also true that in many of the more isolated and rural parts of the country, the crisis is not one of vanishing farmland but vanishing farm economies and vanishing farm communities. The farmland remains, but its farmers are leaving or struggling.

It's no secret that American farming is a disappearing way of life. According to the United States Department of Agriculture (USDA), United States farms declined 8 percent from 2017 to 2024, from 2.04 million farms to 1.88 million, marking the first time the number of United States farms had declined below two million since before the Civil War. Acres of land on farms similarly declined from 900 million acres in 2017 to 876 million acres in 2024.[8]

A big part of the problem is that traditional scales of farming no longer work in the global marketplace, where price, not quality, is the primary economic driver. The USDA reported that net farm income declined in 2023 and 2024, decreasing from $155.9 billion in 2023 to $116.1 billion. Although net farm income was projected to rise slightly in 2025 (when this book was written), overall cash receipts were projected to decline by $1.8 billion and farm labor expenses were projected to rise 3.6 percent to $53.5 billion, a record high. And this was before the disruption of the Donald Trump administration's trade tariffs, immigration policy, and budget cuts upended America's agricultural economy.[9]

Volatilities such as these make it harder and harder for America's small- and medium-scale farmers to compete against their larger competitors. According to USDA, the number of farms in the lowest economic class (defined by summing the sales of agricultural products plus government program payments) dropped the most in the five-year period from 2017 to 2022, by 13 percent (120,970 farms), while those in the next lowest class dropped 5 percent. Meanwhile, farms in the highest economic class (generating $1 million or more) grew by 36 percent (28,566 operations). And these farms in the highest economic class generated 79 percent of the agricultural products sold in America in 2024, up from 69 percent in 2017.[10]

Economies of scale are also creeping into farmland ownership. As land becomes more expensive, independent operators—especially young, minority, and beginning farmers—are finding it harder to buy or lease land, giving advantage to large-scale operations. A growing concern, then, is land owned by non-farm operators, including corporate and foreign investors. According to USDA, 80 percent of all rented American farmland is owned by non-operator owners, i.e. agricultural land-owners not involved in farming.[11] The agricultural industry website AcreTrader reports 11 percent of non-operator-owned farmland is owned by corporations.[12] And based on data collected through 2023, the American Farm Bureau reports that 45.9 million acres of farmland, or 3.6 percent of all privately held United States agricultural land, is owned by foreign entities or individuals, with Canadian investors owning the largest share of that number—12.8 million acres, or 31 percent. Although several recent media stories have focused on China's efforts to acquire American farmland, the Chinese owned only 0.03 percent of American farmland as of 2023.[13]

Indeed, it is getting harder for the individual farmer, or farm family, to acquire farmland, or afford the property taxes and management expenses of owning it. In 2024, the value of United States farmland averaged $4,170 per acre, an increase of 5 percent over 2023 values, or 1.4 percent when adjusted for inflation. From 2018 to 2023, farmland value increased 5.1 percent, or 1.4 percent when adjusted for inflation.

While prices like these may seem reasonable for rapidly urbanizing areas like Chatham County, the price for an acre of farmland varies widely depending on its region in the United States. The average acre of farmland is valued at $4,840 in the Southeast, but it is $8,100 in the Midwest's "Corn Belt" and $7,270 on the Pacific Coast. The type of farmland influences its average price. Between 2023 and 2024, inflation-adjusted average U.S. cropland values rose by 2.2 percent to $5,570 per acre, compared to $1,830 per acre for average pastureland values, a 2.7 percent increase (with all prices adjusted for inflation). The price of renting farmland is also rising, with cropland rental rates increasing 0.8 percent to $160 per acre from 2023 to 2024 and pastureland rental rates increasing by 0.9 percent to $15.5 per acre.[14]

These average prices don't have any relevance in Chatham County, or other parts of America where the "fair market value" of farmland is now based on its value for residential, commercial, or industrial development. If the prevailing per acre value of farmland is something around $15,000 to $30,000, as it is in Chatham County, or more (as it is in even more rapidly developing parts of the country), then those prices just don't make sense for farming, even at an industrial scale. This limits the available farmland to land already owned by existing farmers or passed down through farming families.

But even these land ownership transitions between members of the same family are not secure, for prevailing real estate and estate laws in America make these transfers challenging, especially when there is more than one family heir and the land is encumbered by debt. Although perhaps a cliche, it is nevertheless true that many of today's American farming families are "land rich and cash poor." For many aging farm couples, most of their estate's worth is tied up in the "market" value of their farmland. And in many cases, these couples have children or heirs who are not interested in continuing the farming operations. Or there is one heir who wants to keep farming and other heirs who do not. Too often, this results in the farmland being "liquified," or sold, so its value can be distributed in cash to the heirs. The sad fact is it is easier to divide money than land, so family farms often get sold, even when some family members want to continue farming it.

Farmland Preservation: A "Wicked" Policy Problem

This book is an exploration of efforts to address the "wicked" problem of American farmland loss and conversion through farmland preservation. Although farmland preservation need not be addressed on a policy level—a committed and engaged family is perfectly capable of taking the measures needed to preserve their farmland without any "policy" or outside assistance, except for perhaps an experienced estate attorney—policy does play an increasingly important role in American farmland preservation efforts. Wicked problems are complex issues that require multiple solutions, and a hypothesis of this book is that structural problems influencing the farmland crisis are simply too "wicked" to overcome at scale without thoughtful and committed policy support.

For many, the term "farmland preservation" is synonymous with conservation easements, which are legal documents that restrict or remove the ability to develop a property for residential or commercial purposes. The mechanics and principles of conservation easements will be covered in more detail later in the book. The rise in conservation easements as a tool of farmland preservation corresponds with the popularity of conservation easements in general across America. According to the most recent National Land Trust Census report, a publication of the Land Trust Alliance, the national umbrella organization for America's land trust community, there were 61 million acres conserved through conservation easements in America as of year-end 2020, an increase of more than 15 million acres since the previous census in 2010.[15] Not all that land was farmland—these figures also include conservation easements that protected water quality, biodiversity, created resiliency for natural disasters, and provided other conservation benefits. But a significant percentage of it is. The AFT's Protected Agricultural Lands Database estimates that nearly 7 million acres of agricultural land has been permanently protected, primarily through conservation easements, much of it over the last two decades.[16]

There are good reasons why conservation easements have become America's main farmland preservation tool. One is the incredible land protection success over the last fifty years of America's land trust community, of which I consider myself a proud member. In America, a land trust or land conservancy is a community-based, nonprofit organization that actively works to permanently conserve land. This success has not gone unnoticed by local and state governments, who find conservation easements appealing because they promote voluntary natural resource protection based on the established principles of real estate law, without resorting to intrusive and difficult-to-enforce environmental and land-use regulations. That's why the increase in farmland protection through conservation easements has been accompanied in recent years by the rise of Purchase of Agricultural Conservation Easement (PACE) programs, which are grant programs typically funded by public dollars that help land trusts and other conservation organizations purchase agricultural conservation easements from willing landowners. Again, according to AFT, twenty-nine states had active PACE programs as of 2024, and that's not counting local PACE programs like the one I run in Chatham County.

But conservation easements are not a good fit for every piece of farmland, and far from a perfect solution for every farmland-owning family. As a farmland preservation practitioner, and someone who aspires to be a voice in its policy directions, I fear that conservation easements are being "sold" to farming communities at the expense of more meaningful, though perhaps more challenging, farmland preservation tools. Conservation easements go after the farmland preservation problem from a resource perspective. By intent, they are designed to protect the land and the land's "conservation values," including its future farming potential. But they also create false expectations among a generally uneducated public that we can "fix" farmland loss by focusing exclusively on the land resource issue, saving one farm at a time through conservation easements, when the heart of the matter has as much to do with the changing dynamics of the farming family, and the subsequent sociological crisis in America's farming communities, as the land itself.

The Looming Crisis: Age, Gender, and Race

A big part of America's farmland preservation problem is the aging of its farmers. According to AFT, more than 40 percent of American farmland is owned by farmers aged 65 or older.[17] The 2022 Census of Agriculture reports the average age of the United States farmer is 58.1 years. Although this is only 0.6 years older than the average age from the 2017 census, it is 9.4 years older than the 1945 Census, the first to report an average age. Meanwhile, farmers aged 65 and older increased by 12 percent according to the 2022 Census of Agriculture, even though these farmers owned more than 40 percent of American farmland. And farmers aged 35–64 decreased by 9 percent.[18]

What this aging farmer population means for the future of American farming isn't entirely clear. A 2024 report by the conservative-leaning American Enterprise Institute (AEI) suggests that the Census's average age figure doesn't provide an accurate portrait because it fails to account for diversity across farm sizes and regions. Farm operators in the Midwest, the report notes, tend to be younger, while farms in the South and West tend to have older farm operators. Furthermore, farms in the USDA's "large" and "very large" categories, which together produce 46.1 percent of total agricultural production in the United States, have operators with the lowest average age (57). "Therefore, the largest farms, which are the most crucial for food security and agricultural production, are managed by the youngest farm operators," the reports state.[19]

But the farmland conversion crisis is centered in the small- and medium-sized farms the AEI report appears to discount, and, not coincidentally, the same type of farms that still comprise the "heart" of many of our remaining farm communities. But as these small and mid-size farmers retire or die over the next twenty years, it is expected that around 370 million acres of farmland will change hands. This period of ownership transition is always a precarious moment in the realm of farmland preservation, and the federal or state governments have yet to form any kind of meaningful response to it. Although Congress in the 2018 Farm Bill authorized a ten-member "Commission on Farm Transition," tasked with studying issues impacting the transition of agricultural operations to the next generation of farmers and ranchers, the Commission has not yet been constituted as of 2025.[20]

Sadly, even as the existing American farmer ages, the barriers for new farmers are becoming more daunting. Lack of access to land and financial capital, lack of health insurance and other benefits, childcare expense and considerations, economies of scale, and escalating infrastructure costs are all factors making it difficult for young people who want to enter farming careers or take over farms from previous generations. And the rapidly escalating price of farmland in the type of communities where these new farmers want to live, work, and establish local food systems—often on the fringe of growing urban areas—may well be the greatest barrier of all. The economies of scale in modern agricultural markets exacerbate these barriers, forcing farmers to get big as fast as possible, specialize in crop production, and adopt environmentally damaging production methods. Alternatively, new farmers can face a future where farming will only ever be a part-time job, increasing the stress on family capacities and resources.

A 2022 survey by the National Young Farmers Coalition enumerated the multiple challenges facing young farmers in the twenty-first century. Results indicated that 59 percent of young farmers found buying affordable land very or extremely challenging. For BIPOC (Black, indigenous, or people of color) young farmers, 33 percent found that maintaining access to land was very or extremely challenging, compared to only 17 percent of White young farmers. Forty-one percent of survey respondents listed access to capital as very or extremely challenging, with 54 percent of all BIPOC farmers and 59 percent of Black farmers listing capital access as very or extremely challenging.

Health care expenses, production costs, housing, and student loan debt round out the survey's top challenges for young farmers, with 62 percent of Black young farmers listing student loan debt as a very or extremely challenging barrier, compared to 36 percent of White young farmers. In a section of the survey titled "Barriers to Entry," finding affordable land to buy was listed as the top barrier by 59 percent of survey respondents, with access to capital (41 percent), health care costs (41 percent), cost of production (35 percent), and affordable housing options (33 percent) rounding out the top five.[21]

New farmers not only need access to the physical assets of farming—land, water, equipment, inputs, and labor—they also need cognitive assets such as knowledge and social networks. And the confusing bureaucracies of state, local, and federal farmland assistance agencies only complicate efforts to obtain this knowledge.[22] These barriers are even greater for minority and women farmers. Of the approximately 2 million U.S. farms in 2022, 91 percent were non-Hispanic White, 4 percent were Hispanic, 2 percent were non-Hispanic American Indian or Alaska Native, 1 percent were non-Hispanic Black or African American, and 1 percent were non-Hispanic Asian farms.[23] In 2022, there were 1.2 million female producers in the country, making up 36 percent of the country's farmers, but just 9 percent of U.S. farms are run by women.[24]

American farming has always been explicitly discriminatory in both gender and race. As the author Jessica Shoemaker observes,

> It is no accident that today 98 percent of agricultural lands remained owned and controlled by people who are White. Nor is it accidental that some of the most notorious and persistent regions of concentrated poverty are both rural and racialized.[25]

Recognition of these race and gender barriers, at least, is improving, but as the scale of profitable farming grows ever larger, large-scale farming operations continue to be dominated by White, male farm operators, partners, and stakeholders, leaving women and minorities at the edge of the modern farm economy, often competing over access to market farming outlets such as farmer's markets and community supported agriculture (CSA) cooperatives.

For Shoemaker, these challenges reside in the American property system itself, which she argues is still designed to keep prosperity in racialized patterns. In a legal system designed to support the transfer of property within families as a mechanism for accumulating generational wealth, women and nonwhite farmers start at a disadvantage in their quest to access land. Without this inheritance, new farmers have to acquire land from somewhere else, a process that, by design, too often prevents women and minorities from gaining access to the farming system.[26]

Farming, Social Capital, and Sense of Place

It's no surprise that the twenty-first-century challenges facing American farmers are having a corollary impact on the communities where farming takes place. Between 2010 and 2020, America's rural population declined by 289,000. And while that number may seem relatively small in a nation of 333 million, this is the first time in the history of the United States Census that America's rural population has declined from one decade to the next. For comparison's sake, America's rural population grew 1,516,000 (or 3.4 percent) in the decade between 2000 and 2010.[27] Other population demographics also prove ominous for America's rural, farming communities. During the same 2010 to 2020 period, the rural working-age population (ages 18–65) declined by 4.9 percent, and the rural population under 18 declined by 5.7 percent, while the rural 65 and older population grew by 2.2 percent.[28]

But you don't need demographics to document this decline, especially in America's more remote rural areas. There the desolation of abandoned main streets and decrepit commercial boulevards, vacant homes and untended fields, provide stark visual reminders that many of America's rural communities are struggling, telling the tale more vividly than numbers ever will.

In these isolated rural places, globalization and a rapidly changing culture have led to economic decline—enhancing rural poverty, unemployment, and underemployment—and creating a vicious pattern of emigration to more economically robust areas, typically in America's urban centers. As people and businesses leave rural communities, their school systems suffer. This leads not only to lower levels of educational attainment but also the consolidation or closing of rural schools, which frequently serve as important social centers. Without sustainable career prospects, youth turn to drugs and criminal activity, accelerating the downward curve. Meanwhile, the elderly lack access to affordable, accessible health care, living their final years in isolation or in substandard housing. Churches, locally owned retail and restaurants, and other traditional social institutions of rural America inevitably bear the aftershocks of this systematic social disintegration.[29]

In Chatham County, the problem isn't population decline. We're growing faster than we can handle. Here, increased residential and industrial development, fueled by the growth of the North Carolina Triangle, is pressuring farmers to sell out, even as the rapid escalation in real estate prices is keeping new farmers from moving in. The problem isn't declining Main Streets or decrepit homes, although we have our share, but the new McMansions springing up on the former fields and forests once owned by rural community members.

Either way—decline or growth—these changes inevitably impact what the researcher and author Robert Putnam would call the rural "community," a socially interdependent group of people, participating together in civic, social, and economic life. In his landmark work, *Bowling Alone*, published in 2000, Putnam focused on the civic dimensions of community and its decline, through a primary framework of social capital, a term that Putnam defined as "the connections among individuals' social networks and the norms of reciprocity and trustworthiness that arise from them."[30]

For Putnam, the decline of the American community could be characterized through the decline of social capital. The social capital of shared churches, schools, bowling leagues, civic clubs, and other social institutions makes collective problems easier to solve, Putnam argued, resulting in improved social environments where trust is enhanced through consistent interactions between community members. This mutual and sustained connectivity creates numerous benefits, including the quality of our civic and democratic institutions and their capacity to respond to crisis. Economic systems also improve where social capital is robust, as do health, education, and overall happiness.

Bowling Alone documented a decline in social capital that Putnam traced to the 1970s, when shifting social, economic, and technological dynamics fundamentally changed the fabric of American social life. These included pressures of time and money, mobility and sprawl, television and mass media, and generational differences emerging after World War II. Thus, the decline in social capital is the result of wider sociocultural and economic forces. Nevertheless, for Putnam, enhancing social capital, primarily through enhanced civic participation and reinvigorated democratic frameworks, was his proposed solution to addressing the effects of these negative social forces.[31]

The social capital theories of Putnam and his contemporaries are influenced by the theories of the French sociologist Pierre Bourdieu, who originated the concept of social capital. For Bourdieu, capital had different types—economic, social, cultural, and symbolic. If social capital is a convenient framework for understanding the social dimensions of the farmland conversion crisis, Bourdieu's concept of symbolic capital helps us understand its cultural ones. For Bourdieu, symbolic capital refers to the honor, prestige, or recognition possessed by a group or individual. It is closely related to the concept of symbolic power, which refers to the meanings and power relationships the symbol imposes in the community.[32] Later in this book, I'll use Bourdieu's theories on symbolic capital to help explain the symbolic role farming and farmland play in both the American consciousness and the consciousness of the communities where farmland is being lost and/or preserved.

My analysis of farmland preservation policy and practice is also rooted in the sociological concept of place, which shares social and symbolic dimensions with Bourdieu's capital theory. "Place is a center of meaning constructed by experience," explains the early place scholar Y.F. Tuan, who defined "sense of place" as the meanings and attachment to a setting held by an individual or group:

> The small pleasures and irritations of day-to-day living, the barely registered but omnipresent ambience of sound and smell, the feel of air, soft soil, and hard ground—these are the common experiences of life that may add up to a profound sense of place.[33]

Thus, sense of place is a sensation beyond the processing of sensory and cognitive information, encompassing the ontological process through which humans compose their understanding of the world.

None of the demographics quoted in the first part of this chapter documented the American farmer's decline in social and symbolic capital. None reported on threats to the American farmer's sense of place, or the ontological crisis created in rural communities through the loss of farmland. But these impacts are surely as important to the issue of farmland preservation as escalating real estate prices, industrializing economies of scale, or the aging of the American farmer.

In the chapters that follow, I will spend some time looking at the farmland preservation movement through the perspectives of place and propose preservation frameworks that cast it not only as an effort to preserve natural capital, agricultural land and its broader landscapes, but also social and symbolic capital. This will require a reimagining of farmland preservation beyond its current focus on resource protection to one that encapsulates all the dimensions of place and engages the citizens of a specific place in solutions specifically designed for them, based on their specific needs and challenges. For there is no one farmland preservation crisis in America, just as there is no one type of farm or one farm community. Each is unique in its own way, requiring unique policies and solutions.

Notes

1 "Farms Under Threat 2040: Choosing an Abundant Future," American Farmland Trust, accessed August 9, 2024, https://farmland.org/project/farms-under-threat/.
2 American Farmland Trust, "Farms Under Threat 2040: Choosing an Abundant Future."
3 "Population Growth 2020–2030," North Carolina Office of State Budget and Management, accessed August 9, 2024, https://www.osbm.nc.gov/facts-figures/population-demographics/state-demographer/countystate-population-projections/population-growth-2020-2030.
4 "Storyliving by Disney Announces Asteria, New Residential Community in North Carolina," PR Newswire, December 6, 2023, https://advance-lexis-com.libproxy.clemson.edu/api/document?collection=news&id=urn:contentItem:69T9-R6S1-DXP3-R4HM-00000-00&context=1516831.
5 G. Galin, "Priced Out? The Housing Affordability Crisis in Chatham County," *The Chatham Journal*, June 20, 2024.
6 "2023: The State of Chatham County Report," Chatham County, accessed April 26, 2024, https://www.chathamcountync.gov/home/showpublisheddocument/66008/638373116896130000.
7 "Land For Sale in Chatham County," Land.com, accessed August 12, 2024, https://www.land.com/Chatham-County-NC/.
8 "Farming and Farm Income," USDA Economic Research Service, accessed May 13, 2024, https://www.ers.usda.gov/data-products/ag-and-food-statistics-charting-the-essentials/farming-and-farm-income.
9 "Farm Sector Income & Finances: Farm Sector Income Forecast," USDA, updated February 6, 2016, https://www.ers.usda.gov/topics/farm-economy/farm-sector-income-finances/farm-sector-income-forecast.
10 Daniel Munch, "Over 140,000 Farms Lost in 5 Years," Market Intel, March 7, 2024, https://www.fb.org/market-intel/over-140-000-farms-lost-in-5-years.
11 "Land Use, Land Value & Tenure," USDA Economic Research Service, updated May 8, 2025, https://www.ers.usda.gov/topics/farm-economy/land-use-land-value-tenure/farmland-ownership-and-tenure#:~:text=The%20majority%20of%20rented%20acres,as%20%22operator%20landlords%22).
12 "Who Owns U.S. Farmland?" AcreTrader, November 29, 2023, https://acretrader.com/learn/farmland-investing/who-owns-farmland.
13 "Foreign Footprints: Trends in U.S. Agricultural Land Ownership," American Farm Bureau, January 14, 2025, https://www.fb.org/market-intel/foreign-footprints-trends-in-u-s-agricultural-land-ownership.
14 "Land Use, Land Value & Tenure," USDA Economic Research Service, updated May 8, 2025, https://www.ers.usda.gov/topics/farm-economy/land-use-land-value-tenure/farmland-value.
15 "61 Million Acres Voluntarily Conserved in America," Land Trust Alliance, December 2, 2021, https://landtrustalliance.org/newsroom/press-releases/61-million-acres-voluntarily-conserved-in-america-2020-national-land-trust-census-report-reveals.
16 "Protected Agricultural Lands Database," AFT Farmland Information Center, accessed August 19, 2024, https://farmlandinfo.org/statistics/pald/.
17 "2021 Transition Recommendations," American Farmland Trust, accessed August 14, 2024, https://farmland.org/transition2021/.
18 "2022 Census of Agriculture Impacts the Next Generations of Farmers," USDA, accessed August 12, 2024, https://www.ers.usda.gov/topics/farm-economy/land-use-land-value-tenure/farmland-value/https://www.usda.gov/media/blog/2023/02/22/2022-census-agriculture-impacts-next-generations-farmers.

19 Eric Belasco and Joseph Glauber, "An Aging Farm Population: Cause for Concern?" American Enterprise Institute, July 2024, p. 4, https://www.aei.org/research-products/report/an-aging-farm-population-cause-for-concern/.
20 "2021 Transition Recommendations," American Farmland Trust, accessed August 14, 2024, https://farmland.org/transition2021/.
21 "Building a Future with Farmers 2022: National Young Farmer Survey," National Young Farmers Coalition, accessed August 15, 2024, https://www.youngfarmers.org/wp-content/uploads/2022/09/NationalSurveyReport22-1.pdf.
22 Liz Carlisle et al., "Securing the Future of U.S. Agriculture: The Case for Investing in New Entry Sustainable Farmers," *Elementa: Science of the Anthropocene* 7 (2019): 17.
23 "Understanding Farm Diversity: Insights from the Agricultural Resource Management Survey," USDA Economic Research Service, May 30, 2024, https://www.ers.usda.gov/amber-waves/2024/may/understanding-farm-diversity-insights-from-the-agricultural-resource-management-survey/#:~:text=Of%20the%20approximately%202%20million,were%20non%2DHispanic%20Asian%20farms.
24 "Honoring Female Farmers and Ranchers this Women's History Month," USDA, March 15, 2023, https://www.usda.gov/media/blog/2023/03/15/honoring-female-farmers-and-ranchers-womens-history-month.
25 Jessica A. Shoemaker, "Fee Simple Failures: Rural Landscapes and Race," *Michigan Law Review* 119, no 8 (2021): 1699.
26 Shoemaker, "Fee Simple Failures," 1723.
27 Kenneth Johnson, "Rural America Lost Population Over the Past Decade for the First Time in History," University of New Hampshire Carsey School of Public Policy, February 22, 2022, https://carsey.unh.edu/publication/rural-america-lost-population-over-past-decade-first-time-history#:~:text=Between%202010%20and%202020%2C%20the%20rural%20population%20declined%20by%20289%2C000,of%20221%2C000%20(0.5%20percent).
28 James Davis et al., "Rural America at a Glance: 2022 Edition," USDA Economic Research Service, accessed August 15, 2024, https://www.ers.usda.gov/webdocs/publications/105155/eib-246.pdf.
29 Shelley Clark, Sam Harper, and Bruce Weber, "Growing Up in Rural America," *RSF: The Russell Sage Foundation Journal of the Social Sciences* 8, no. 4 (2022): 1–47.
30 Robert D. Putnam, *Bowling Alone: The Collapse and Revival of American Community* (Simon & Schuster, 2000), 19.
31 Putnam, *Bowling Alone*, 19.
32 Pierre Bourdieu, "Structures, Habitus, Power: Basis for a Theory of Symbolic Power," *Culture/Power/History: A reader in Contemporary Social Theory* 155 (1994).
33 Yi-Fu Tuan, "Place: An Experiential Perspective," *Geographical Review* 65, no. 2 (1975): 152, 164.

Bibliography

American Farmland Trust. "Farms Under Threat: Choosing an Abundant Future." Accessed August 9, 2024. https://farmland.org/project/farms-under-threat.

American Farmland Trust Farmland Information Center. "Protected Agricultural Lands Database." Accessed August 19, 2024. https://farmlandinfo.org/statistics/pald/.

Belasco, Eric, and Joseph Glauber. "An Aging Farm Population: Cause for Concern?" American Enterprise Institute. July2024. https://www.aei.org/research-products/report/an-aging-farm-population-cause-for-concern/.

Bourdieu, Pierre. 1994. "Structures, Habitus, Power: Basis for a Theory of Symbolic Power." In *Culture/Power/History: A Reader in Contemporary Social Theory*, edited by Nicholas B. Dirks, Geoff Eley, and Sherry B. Ortner. Princeton University Press.

Carlisle, Liz, 2019. "Securing the Future of U.S. Agriculture: The Case for Investing in New Entry Sustainable Farmers," *Elementa: Science of the Anthropocene* 7. doi:10.1525/elementa.356.

Clark, Shelley, Sam Harper, and Bruce Weber. 2022. "Growing Up in Rural America," *RSF: The Russell Sage Foundation Journal of the Social Sciences* 8, no. 4. doi:10.7758/rsf.2022.8.3.01.

Munch, Daniel. "Foreign Footprints: Trends in U.S. Agricultural Land Ownership." American Farm Bureau. January 14, 2025. https://www.fb.org/market-intel/foreign-footprints-trends-in-u-s-agricultural-land-ownership.

United States Department of Agriculture. "Farm Sector Income & Finances: Farm Sector Income Forecast." Updated February 6, 2016. https://www.ers.usda.gov/topics/farm-economy/farm-sector-income-finances/farm-sector-income-forecast.

United States Department of Agriculture. "2022 Census of Agriculture Impacts the Next Generations of Farmers." Accessed August 12, 2024. https://www.ers.usda.gov/topics/farm-economy/land-use-land-value-tenure/farmland-value/https://www.usda.gov/media/blog/2023/02/22/2022-census-agriculture-impacts-next-generations-farmers.

USDA Economic Research Service. "Farm Sector Income & Finances: Farm Sector Income Forecast." Updated February 6, 2016. https://www.ers.usda.gov/topics/farm-economy/farm-sector-income-finances/farm-sector-income-forecast.

USDA Economic Research Service. "Land Use, Land Value & Tenure." Updated May 8, 2025. https://www.ers.usda.gov/topics/farm-economy/land-use-land-value-tenure/farmland-value.

National Young Farmers Coalition. "Building a Future with Farmers 2022: National Young Farmer Survey." Accessed August 15, 2024. https://www.youngfarmers.org/wp-content/uploads/2022/09/NationalSurveyReport22-1.pdf.

Putnam, Robert D. 2000. *Bowling Alone: The Collapse and Revival of American Community*. Simon & Schuster.

Shoemaker, Jessica A. 2021. "Fee Simple Failures: Rural Landscapes and Race," *Michigan Law Review* 119, no 8. doi:10.36644/mlr.119.8.fee.

Tuan, Yi-Fu. 1975. "Place: An Experiential Perspective," *Geographical Review* 65, no. 2.

3 Setting the Stage for Crisis

American Farmland Preservation Policy in the Nineteenth and Twentieth Centuries

In his first major speech on farm policy, Franklin Delano Roosevelt addressed the land-use issues that were plaguing American agriculture. Speaking on the campaign trail in Topeka, Kansas, on September 14, 1932, Roosevelt called for a vigorous "interdependence" of approaches, insisting that longer perspectives would be necessary to address the many ills facing Depression-era farmers.

> We must have, I assert with all possible emphasis, national planning in agriculture ... we must avoid the present tendency to jump from one temporary expedient to another. We need unity of planning, coherence in our Administration and emphasis upon cures rather than drugs.

In tying America's farm crisis to concepts such as planning, cures, and coherence, Roosevelt was advocating for a brand of "New Conservation" that had become popular among the Progressives of the 1920s. Supported by a coalition of eastern elites and agronomists working within the United States Department of Agriculture (USDA), this outlook believed the poverty and social degradation threatening the viability of rural communities could be addressed through regional planning for land and water resources.[1]

Roosevelt's speech outlined two objectives for the New Conservation's vision of agricultural reform. First, he would order the USDA to institute a "definite policy looking to the planned use of land." Second, this new planning infrastructure would conduct a soil survey:

> to be directed toward the problems of proper utilization of the land ... It should lead to mapping and classification of land of all kinds, to determine which lands are best suited for agricultural production, which lands are marginal and which lands are suited only to growing tree crops.[2]

The New Conservation sought to address the widespread exploitation of soils and prime farmland characterizing the latter stages of Manifest Destiny in the Great Plains and American West, as well as the systematic poverty that still plagued the South and other agricultural regions of the country. Fueled by

DOI: 10.4324/9781003644545-4

the manpower shortages of the Civil War, the second half of the nineteenth century saw transformational changes in America's agricultural technology. The development of mechanical reapers and threshers, seed drills, harrows, rollers, horse rakes, and hay loaders all emerged out of this period.

These technological improvements expanded the scale of American agricultural production. In Iowa, for example, wheat production grew 250 percent between 1860 and 1870, rising from 8.4 million bushels in 1860 to 29.4 million bushels a decade later. In 1870, the newly formed USDA estimated that American farm worker productivity had increased 13 percent since 1860, compared with an 8 percent increase from the previous decade. Other war-time innovations opened new farming opportunities. In 1862, America's war-time Congress passed the Homestead Act, encouraging the agricultural settlement of the Great Plains. Passed the same year, the Pacific Railway Act, which spurred the connection of both American coasts via rail, also opened agricultural settlement in previously unfarmed territories of the American landscape. In the 1870s, 190 million new acres were put under cultivation, and by the 1880s, agricultural was moving onto the West's arid plains.[3]

But these seemingly boundless opportunities for American farming bore ill omens for its soil and productive farmland. As pioneers and homesteaders pushed to the United States' borders, they cultivated more and more acreage in fragile ecological landscapes. Subsequently, forests, soil, and wildlife rapidly disappeared from vast swaths of America, as farmers, hunters, foresters, and miners stripped local land before moving on to virgin territories.

In the South, the plantation system's highly concentrated practice of land ownership went largely unchanged after the Civil War. Here the problem was not the exploitation of new lands but the increasingly unprofitable farming of lands already depleted by decades of exploitative practices. Cotton was still king in the American South, but the problem became finding laborers to farm it. In response, landowners turned to sharecropping. In return for access to land, the sharecropper typically received half the annual crop yield as payment. The landowner usually provided the land, livestock, and farming implements as part of the contract agreement, but other necessities (groceries, clothing, household supplies, seed, fertilizer, etc.) were supplied on credit using the sharecropper's yield as collateral.

For landowners and merchants, this crop-lien system controlled labor and minimized risk, providing control over the sharecroppers' market for both inputs and outputs.[4] From an ecological standpoint, sharecroppers squeezed more and more production from increasingly depleted land, usually with little attention to regenerative or conservation practices. Socially and economically, the sharecropper was chained to the land by debt and poverty, with little opportunity for escape. An exploitative, amoral system, it endured until the price collapses of the 1920s hit the cotton market, finally ruining the South's landowners just as its sharecroppers had been ruined for decades.

Unsustainable farming practices also plagued the newer farming communities of the Great Plains and West Coast. To enhance commodity crop production in these arid regions, farmers turned to "dryland" farming practices, relying on scientifically unsound practices like shallow tilling to retain soil moisture without irrigation. Although dryland farming enabled farmers to produce wheat and other drought-resistant crops in arid regions, it took a heavy toll on soil fertility.[5] Similarly, great swaths of western timber were harvested unsustainably, with little attention to soil preservation or restoration. "The relationship between profitable yields and the balance of nutrients in soil emerged as one of the most intractable problems in American agriculture before the twentieth century," observes historian Steve Stoll. "Though farmers understood that fertility could be taken, few understood the best way to put it back and almost no one knew the chemistry behind any process."[6]

Rise of the Agrarian Movement

If the agricultural boom of the 1870s and 1880s was fueled by technological advances and the expanded horizons of manifest destiny, its proliferation undermined its economic success. Although a third of Americans worked in agriculture by the end of the nineteenth century, overproduction led to falling prices and plunging farm incomes. By 1897, the price of corn had fallen to 30 cents a bushel, down from 41 cents a bushel in 1874; in 1871, Kansas wheat sold for an average of $1.56 a bushel but dropped to $1.27 a bushel by 1891.[7]

Historian James I. Stewart attributes the farming crisis of the 1890s to three factors, each contributing to the growing risks of agricultural production following the Civil War. First was the farmers' growing dependence on creditors, merchants, and railroads, creating a sense of vulnerability and servitude among a farmer class long prideful of their self-sufficiency. Second was the global integration of grain markets, undermining American dominance and increasing price volatility. Third were risky and wasteful farming practices, easily undone by drought and other environmental calamities.

In response to this crisis came the initial wave of federal and state laws aimed at supporting America's agricultural economy. First came the "Granger" laws of the 1860s and 1870s (the word "granger" derived from English slang for "farmer," originating from the Latin word for grain, *granum*). Enacted by midwestern states to regulate rapidly rising railroad and grain elevator fees, these Granger laws were quickly challenged in court by their powerful adversaries in the railroad industry, leading to a Supreme Court rulings that gave the federal government sole regulation power over interstate commerce. Congress responded with the Interstate Commerce Act of 1887, giving the federal government oversight over long-distance rail shipping, and the Sherman Antitrust Act of 1890, prohibiting monopolies and other corporate restraints on trade.[8]

From this political, economic, psychological, and legal unrest emerged the agrarian movement, a political crusade of the late nineteenth and early twentieth century's Progressive Era. The agrarian movement's leaders and organizations attempted to improve the economic status of farmers through land reforms such as decentralized land ownership and price supports. "Agrarianism blamed the farm crisis on an imbalance between the agricultural and industrial sectors, to be readjusted by legislation specifically benefitting the farmers," writes the historian Kenneth Finegold. "This policy was tied to an ideology of agricultural fundamentalism, which held that farming was the ultimate source of all wealth and hence more basic, virtuous, and democratic than other occupations."[9]

This agricultural fundamentalism was famously embodied in Thomas Jefferson's description of the American farmer in his 1785 book, *Notes on the State of Virginia*:

> Those who labor in the earth are the chosen people of God, if ever he had a chosen people, whose breasts he has made his peculiar deposit for substantial and genuine virtue. It is the focus in which he keeps alive that sacred fire, which otherwise might escape from the face of the earth. Corruption of morals in the mass of cultivators is a phenomenon of which no age nor nation has furnished as an example. It is the mark set on those, who, not looking up to heaven, to their own soil and industry, as does the husbandman, for their substance, depend for it on casualties and caprice of customers. Dependence begets subservience and venality, suffocates the germ of virtue, and prepares fit tools for the design of ambition.[10]

Jefferson's philosophical musings on the social, spiritual, and moral role of the American farmer soon evolved into the bare-knuckled populism of Andrew Jackson, pitting eastern cities and financial centers against western settlers. This Jacksonian democracy was largely a Western agrarian movement whose intent was to break the power of the Northeast's privileged aristocracy while preparing the way for more egalitarian forms of government based on majority rule and economic individualism. These same divisions—rural versus urban and farming versus industry—fueled the tensions leading to the Civil War, pitting the agrarian American South versus the industrializing American North.[11]

However, the converging political interests of the agrarian South and the agrarian Northwest soon reunited in the National Grange, America's first successful national agricultural organization. Formed in 1867, the Grange's original mission was social and fraternal, borrowing some of its rituals from the Freemasons. The Grange's magic formula combined economic self-organization, cooperation, and mutual aid with social outings and a sense of shared community for American farmers.[12]

By 1873, Grange membership had grown to nearly 700,000 members in 9,000 chapters across the United States, and its focus soon turned political.

Grange members naturally gravitated to the agrarian movement, calling for the kind of state control and economic regulation soon established in the "Granger" laws. By the late nineteenth and early twentieth century, the Grange's version of agrarian populism was adopted by national politicians eager to garner the political support of its members, notably that of three-time Democratic presidential nominee William Jennings Bryan, who based his political platform on the superiority of farm life over city life. Rural America, Bryan argued, was not only more healthful than urban American, but also more generous and hospitable, and produced more informed and independent voters.[13]

The Great Depression and the New Deal

The political rhetoric of Bryan and his counterparts did little to alleviate the plight of American farming, which was in crisis again by the late 1920s. World War I had brought another boom cycle, doubling farm incomes between 1916 and 1919. But incomes fell just as dramatically after 1920, when the end of the war returned stability to European markets and stymied global demand for American farm products. If other sectors of the American economy flourished in the "Roaring '20s," American farmers once again found themselves caught between crushing consumer inflation and falling commodity prices. Overproduction exacerbated this vise, with many commercial farmers chasing larger yields to collateralize risky loans on land and machinery. Meanwhile, tenants and sharecroppers remained stranded on marginal lands with little hope for anything resembling economic prosperity.[14]

Adopting the progressive movement's anti-capitalism rhetoric, Agrarians blamed this crisis on policy imbalances between the agricultural and industrial sectors, calling for legislative fixes. In reality, the decline was due to a more complex framework of political, economic, social, and environmental issues. When the 1929 stock market crash ignited the Great Depression, these factors collided into a full-blown collapse of America's agricultural economy. Meanwhile, decades of natural resource exploitation and poor farming practices, exacerbated by drought, resulted in the Dust Bowl's ecological catastrophe. In 1934, twenty-two serious dust storms ravaged the Great Plains, followed by forty in 1935, sixty-eight in 1936, and seventy-two in 1937—an unprecedented period of ecological disaster not experienced before or since.[15] If some of the more humid regions of the Great Plains escaped the dust, they could not escape the drought. Some parts of the American Midwest received no rain at all in 1931, and 1934 proved to be the driest recorded year in the region's history until that time. No wheat at all was harvested in Thomas County, Kansas, in 1933, 1935, 1936, and 1940. By 1935, the United States was forced to import wheat from other countries for the first time.[16]

Even before the Dust Bowl, agricultural policymakers within the federal government were looking at land use as a cause of the agricultural crisis.

Prominent among them was the agricultural economist L.C. Gray, who wanted to extend scientific forestry management principles to agricultural lands. Trained at the University of Wisconsin, Gray had been appointed head of the newly created Division of Land Economics within the USDA in 1919. Two years later, Gray became chair of USDA's Committee of Land Utilization, which published its initial "Yearbook of Agriculture" in 1923. Hailed as the first comprehensive, data-based study of America's agricultural land-use issues, Gray's "Yearbook" provided a systematic and devastating analysis of America's deteriorating soil resources, depicting nineteenth-century efforts to hurry public land into private ownership as the main culprit.

The Yearbook's evaluation highlighted Malthusian characterizations of America's declining agricultural system, predicting catastrophic outcomes for the country's growing population, and framing the loss of agricultural lands as a national security issue. For Gray, wasteful land use and an unregulated market were largely to blame for the unsustainable forestry and soil practices that threatened the "entire fabric of our rural civilization."[17] Thus, Gray's analysis echoed themes of the earlier agrarian movement while anticipating the systemic ecological issues of the Great Depression, bridging the conservation movements of the Progressive and New Deal eras.

Meanwhile, in 1928, a soil surveyor from North Carolina named Hugh Hammond Bennett published a highly influential USDA bulletin titled "Soil Erosion: A National Menace." A graduate of the University of North Carolina, Bennett began his career as a soil surveyor for the USDA, where he became convinced the problem of soil erosion not only threatened individual farms but also rural economies. Bennett's bulletin called for a comprehensive approach to erosion control based on the premise that soil should be used according to its natural characteristics. "As a Nation we are doing very little to abate the evil effects of erosion," Bennet wrote.

> Everyone who knows anything about it admits the problem is a serious one, but few realize how very devastating is the wholesale operation of erosion. There is necessity for a tremendous national awakening to the need for action in bettering our agricultural practices in this connection, and the need is immediate.[18]

But the warnings of Gray and Bennett had gone largely unheeded by President Herbert Hoover's administration, which depicted the farm crisis as a management issue, the result of inefficient production and distribution, not ecological jeopardy. "There is no relief to the farmer by extending Government bureaucracy to control his production and thus curtail his liberties, nor by subsidies that only bring more bureaucracy and ultimate collapse," Hoover declared at the Republican Convention of 1932.[19]

Presidential candidate Franklin Delano Roosevelt, on the other hand, was happy to place the blame for the American farm crisis squarely on the big business interests and laissez-faire policies of his Republican predecessors—Hoover (1929–1933), obviously, but also the earlier administrations of Warren G. Harding (1921–1923) and Calvin Coolidge (1923–1929). Stepping boldly into the tradition of political agrarianism, Roosevelt campaigned on a variety of agrarian policy proposals, ranging from price fixing to tariff protections.[20]

Once in office, however, Roosevelt's agricultural initiatives largely turned toward the production controls and incentivized conservation management programs that still dominate Federal farm policy. But Roosevelt remained a big thinker, influenced by the "big" ideas of others. The comprehensive land-use reforms of the Progressive era still captured his imagination, and he didn't abandon these ideas completely, at least in his early administrations.

The influence of L.C. Gray's 1923 "Yearbook" report was on display when Roosevelt's newly formed Natural Resources Board (NRB) published a comprehensive natural resource inventory in 1934. "The quality of the analysis surpassed anything previous to it; the report reads remarkably well even today. It anticipated many of the concerns of the 1970s environmental movement and suggested solutions that remain untried and scarcely considered," Lehman observes.[21] Emphasizing public interests over private property rights, the inventory spurred an ambitious land-use reform agenda that lasted the rest of the decade, even if it was largely unsuccessful. Later renamed the Natural Resources Planning Board (NRPB), the NRB called for the purchase of approximately 75 million acres of private farmland to retire it from wasteful agricultural practices.

Inspired by Roosevelt's Progressive influences, New Deal reformers such as Hugh Hammond Bennett similarly called for a new administration guiding national soil conservation. In 1935, Bennet's fervent, moralistic appeals on soil conservation were rewarded with the establishment of the USDA's Soil Conservation Service (SCS), where he served as its chief until 1951. In 1994, Bennet's SCS was renamed the Natural Resource Conservation Service (NRCS). Today, NRCS remains the agency responsible for administration and oversight of federal conservation programs, many of which originated out of the New Deal. Yet Bennett understood the broader success of his plan hinged on establishing local administrations. The American farmer, he realized, would be suspicious of federal government's coercive soil management enforcement. His solution was to establish local soil conservation districts, leveraging local leadership to implement federal agriculture policy at local levels.

America's first local soil conservation district was established in 1937 in Bennett's North Carolina home. By 1942, there were 771 local soil conservation districts in forty-one states, covering nearly 452 million acres. Today there are approximately 3,000 conservation districts covering almost every county in America, with more than 17,000 citizens serving on their local governing boards.[22] In its early years, Bennett's SCS gradually introduced proposals to implement land-use planning and regulation into the local districts' supervisory roles. However, by the end of the 1960s only a few districts in the American West had implemented any type of land-use regulation.

Another early effort at large-scale agricultural land-use planning was the Land Utilization Program (LUP) and affiliated efforts, a federal program that purchased large tracts of eroded and submarginal land, converting them to more passive uses such as forestry, grazing, wildlife, and recreation areas. The program was supported by Gray and other progressives within USDA, including Rexford Tugwell, the former Columbia University professor who had emerged as an important agricultural advisor to Roosevelt. By 1934, Tugwell was serving as an undersecretary at USDA, where he had helped create the Agricultural Adjustment Administration (AAA), which paid farmers to voluntarily reduce their production.

Gray, Tugwell, and others within Roosevelt's administration viewed poverty and soil erosion caused by misguided settlement patterns and improper land use as the root of the farming crisis. The LUP's agenda, therefore, expanded beyond the acquisition and retirement of depleted land to include broader social, economic, and environmental initiatives. Farming alone could not revitalize rural America, these New Deal thinkers argued, so the program tried to develop broader opportunities for agricultural economies.

In its broader social objectives, the LUP never achieved much success, but its land acquisition initiatives were more impactful. By 1946, when Congress halted the program, 11 million acres had been acquired and retired from unsustainable farming practices. Though this was far short of the 75 million originally envisioned, the program acquired and converted depleted farmland in almost every American state. In North Carolina, for example, 15,000 acres of worn-out cotton land were transformed into the William B. Umstead State Park. On the western plains, the Cimarron, Kiowa, and Comanche National Grasslands all began as LUP projects.[23]

But ultimately the land-use reforms of the SCS and LUP failed to achieve the widespread social and ecological benefits their creators envisioned. Lehman attributes these failures to three factors. First was the political deadlock of the latter New Deal Era. By the 1940 presidential campaign, land-use reform had become an unpopular political issue, losing Roosevelt's support. Second, implementation of land-use regulation was weakened by local conservation districts' tendency to compromise with local farmers, weakening the impact of the SCS's national policy.

But the most significant reason for these failures, Lehman argues, was the SCS's institutional rivalry with the network of extension services at land grant colleges. The powerful Farm Bureau generally sided with the more conservative extension services, and in 1942 lobbied for the abolishment of the SCS. Under these pressures, the beleaguered SCS abandoned its more ambitious visions of land-use control but continued to provide voluntary erosion control assistance to farmers.[24]

Rise of the Green Movement

In the decades following World War II, American agriculture and agricultural policy changed dramatically. As large-scale "agribusinesses" began to replace small family farms, the laissez-faire agricultural policies of the Eisenhower and Nixon administrations replaced the reform movements of the Progressive and New Deal eras. This post-war period saw another leap in agricultural technology, as new advances in seeds, fertilizers, pesticides, and machinery resulted in greater efficiencies.

Malthusian warnings about a starving American population diminished as production increased, and the conservation ethic that animated 1930s agricultural policy gave ground to policies favoring commodification and policies of scale. Not surprisingly, big agriculture was unenthusiastic about federal farmland protection measures, and its lobbyists united successfully to oppose federal land-use legislation in the 1970s.[25]

But as the agricultural mainstream changed focus from conservation practices to industrial production, a grassroots environmental movement emerged to challenge its harmful land use practices. And a University of Wisconsin professor named Aldo Leopold, who had started his career in the U.S. Forest Service, emerged as one of this movement's most important early voices, despite having died in 1948.

Leopold's influential book was titled *A Sand County Almanac: And Sketches Here and There,* published in 1949, a year after his death. At the root of Leopold's writing was a frustrated admission that the social and ecological objectives of the New Conservation movement had failed to achieve a meaningful reckoning with ethical conservation and land-use practices. This failure, Leopold argued, led to destructive management practices and an agricultural system that regarded land only for its economic value, not its social, ecological, and community ones.

Leopold famously concluded *A Sand County Almanac* by calling for a new "land ethic," enlarging "the boundaries of the community to include soils, waters, plants, and animals, or collectively: the land." Noting that "no important change in ethics was ever accomplished without an internal change in our intellectual emphasis, loyalties, affections, and convictions," Leopold's vision for a land ethic "changes the role of Homo sapiens from conqueror of the land community to plain member and citizen of it. It implies respect for his fellow-members and also respect for the community as such."[26]

It was, perhaps, a familiar notion, not all that different from the political agrarianism of earlier decades, but something about Leopold's literary eloquence took hold of America's emerging "Green" consciousness with an impact that earlier, policy-driven approaches lacked. "Within two decades Leopold's manifesto became the intellectual touchstone for the most far-reaching environmental movement in the United States," writes Roderick Frazier Nash. "Few today would challenge Aldo Leopold's reputation as one of the seminal thinkers in the modern American development of environmental ethics."[27] And Leopold's "land ethic" concept influenced other important books of the early environmental movement, including Rachel Carson's *Silent Spring* (published in 1962) and James Lovelock's *Gaia: A New Look at Life on Earth* (published in 1979).

Farm Crisis in the 1980s and the Modern Farmland Preservation Movement

If Leopold's land ethic took hold in a new generation of agrarians and environmentalists, it did little to address the continuing economic weaknesses of the American agricultural system. By the 1970s, people all over the world were eating more bread, meat, and poultry—resulting in increasing demand for American grains like corn, soybeans, and wheat. In 1972, the Soviet Union disrupted the international grain market by making secret deals with America's five biggest grain companies worth almost $1.5 billion (roughly $10.5 billion in today's dollars). In response to this rapid increase in global demand, U.S. grain reserves were lowered, again raising prices, and Nixon Agricultural Secretary Earl Butz encouraged U.S. farmers to "plant fence row to fence row," advising farmers to "get big or get out … adapt or die."

Under Butz and the Nixon administration, agricultural progress was once again measured by increasing yields, with little attention paid to the harmful health, economic, or environmental effects of industrial agriculture. With demand skyrocketing, land was at a premium. Farmers drained wetlands, cleared forests, and tore out conservation windbreaks to add productive acreage, often with direct assistance from USDA's Soil Conservation Service.[28]

Property values for farmland also increased dramatically. Between 1969 and 1978, the value of U.S. farmland increased 73 percent. Farmers leveraged these assets with risky loans to boost production, chasing higher paydays. But disaster was looming in the form of ballooning inflation, exacerbated by the Arab oil embargo of the late 1970s and the collapse of the Russian grain market, which President Jimmy Carter ordered embargoed in 1980 after the Soviet Union's invasion of Afghanistan. Global median inflation rates that were 5.5 percent in 1970 rose to 14 percent by 1980 before topping at 21.5 percent in 1981. By 1986, American agricultural exports had fallen 50 percent from their peak in 1981. And as agriculture markets faltered, the aggressive financing that took place in the 1970s now haunted both lenders and farmers. The combination of falling farm incomes and higher interest rates plummeted asset values. In both 1985 and 1986, sixty-eight agricultural banks and lending institutions collapsed, followed by fifty-eight more in 1987. By the end of the decade, an estimated 300,000 family farmers had defaulted on loans, and more than 250,000 farmers went out of business.[29]

Cratering demand, skyrocketing interest rates, and bankrupt farmers hardly seems a recipe for a burgeoning farmland preservation movement, but other forces were at play. One was rapid urbanization. USDA estimates that four percent of agricultural land became newly subjected to urban influence between 1980 and 2000, including 13 percent of farmland in Appalachia and the Southeast. Historian David Goldfield calls this the emergence of the multi-centered metropolis, characterized by concentrated suburbanization on the metropolitan periphery.[30]

Another important influence was the confluence of the agrarian and environmental movements, influenced by "New Agrarian" writers such as Wendell Berry and grassroots publications such as 1977's "Progress as If Survival Mattered: Handbook for a Conserver Society." This popular guide argued that the protection of prime farmland from urban development was necessary for ecologically sustainable agriculture. Also influential was the rise of the conservation easement, which allowed nonprofit "land trusts" to permanently restrict development on sensitive natural and agricultural areas. In 1981, a coalition including the Sierra Club, National Resources Defense Council, National Wildlife Federation, and the Sierra Club formed the American Farmland Trust, the first national nonprofit land trust focused on the issue of farmland preservation.[31]

Although it emerged out of the intellectual and social strains of the environmental movement, the modern farmland preservation movement preferred the neo-Malthusian rhetoric of resource crisis, focusing on the specter of urbanization rather than improved conservation management. Scholar Michael Bunce wrote:

> At its height in the mid-1970s and mid-1980s, the farmland preservation discourse was dominated by the language of resource scarcity. That the arguments were simplistic and alarming made them all the more persuasive, especially in the molding of public opinion.[32]

Aldo Leopold would have no doubt been disappointed. Yet again, his plea for a "land ethic" took a backseat to the politically palatable agrarianism of resource and food supply protection, even if these arguments could make no direct statistical link between farmland conversion and declining food production. And for the time being, at least, these positions also mostly ignored the social and economic travails hammering rural communities. After all, a bankrupt farm family hardly needs the environmental protections of a conservation easement, especially if they must sell their land at something resembling market price to escape their debts. Indeed, between the eve of the farm crisis in 1978 and 1992, 15 percent of all farms in the United States disappeared, with higher numbers in Iowa and other agricultural states.[33] Nominal farm income (unadjusted for inflation) peaked in 1973 at $34.4 billion and fell to a record low of $14.3 billion in 1983, a 58 percent decline. Farmland prices also suffered, with some declining 60 percent.[34]

With economics like this, it's no wonder farm families were feeling pessimistic about the future of American agriculture by the end of the 1980s. In 1991, *Successful Farming* magazine conducted a poll of American farmers that was a duplicate of a 1974 poll. The results quantified the economic turmoil facing American farm families: farmers who hoped to have their children take over their farm operations declined from 68 percent in 1974 to 55 percent in 1991. Fewer farmers felt successful—80 percent compared to 92 percent. And more worried about their future—56 percent compared to 34 percent.[35] "The experiences of the previous decade had reinforced the idea that bad things happen to good people. People who once believed that hard work could take them anywhere no longer believed it," observes sociologist Pamela Riney-Kehrberg. "Even as times improved, people continued to feel uneasy, with every setback and difficulty reminding them of the hard times they had suffered."[36]

While the economic situation gradually improved in the 1990s and 2000s, the impacts of the 1980s farm crisis remain. In her study of Iowa, Riney-Kehrberg reports that, in 1978, the average Iowa farmer was 47.2 years old; by 1992 the average age increased to 50.3. As age increased, crop diversification declined, with more Iowa farmers focused on the relatively stable commodities of corn and soybeans, while livestock farmers increasingly turned to industrial-style production in concentrated animal feeding operations.[37]

Farm communities also changed. As farmers went bankrupt, young people turned away from farming, and community institutions such as schools, stores, and hospitals suffered. According to a report by the National Center for Education Statistics, between 1986 and 1994, 415 schools in small, rural school districts shut their doors, and there were as many closures of small rural school districts as in all other categories of school districts combined.[38] "As community symbols, the schools staying open or closing is one of the most telling markers of how people think the community is doing," writes Robert Wuthnow in his study of rural America.[39]

Many of these social problems have gotten worse, not better, in the intervening years. Small and medium-sized farms continue to decline in overall numbers, and the average American farmer is still getting older, even as the barriers to farmland access become more and more structurally challenging for young and minority farmers. Add to the closing of schools, stores, and factories the opioid and fentanyl crisis of the early twenty-first century, along with the migration of rural youth to urban centers, and you have a recipe for continued social decline, with little hope for a return to rural prosperity. Hanging these hopes on the mantle of farmland preservation hardly seems like a formula for success, but let us not underestimate the symbolic capital of American farmers and their land.

Notes

1 Sarah T. Phillips, *This Land, This Nation: Conservation, Rural America, and the New Deal* (Cambridge University Press, 2007), 22.

2 Franklin D. Roosevelt, "Campaign Address in Topeka, Kansas, on the Farm Problem, September 14, 1932," The American Presidency Project, accessed November 3, 2024, https://www.presidency.ucsb.edu/documents/campaign-address-topeka-kansas-the-farm-problem.

3 Wayne D. Rasmussen, "The Civil War: A Catalyst of Agricultural Revolution," *Agricultural History* 39, no. 4 (1965): 187–195.
4 Max J. Pfeffer, "Social Origins of Three Systems of Farm Production in the United States," *Rural Sociology* 48, no. 4 (1983): 553.
5 "Dryland Farming," Encyclopedia of the Great Plains, accessed February 3, 2025, http://plainshumanities.unl.edu/encyclopedia/doc/egp.ag.027.
6 Steven Stoll, *Larding the Lean Earth: Soil and Society in Nineteenth-Century America* (Hill & Wang, 2003), 17.
7 C. Knick Harley, "Western Settlement and the Price of Wheat, 1872–1913," *The Journal of Economic History* 38, no. 4 (1978): 865–878.
8 James I. Stewart, "The Economics of American Farm Unrest, 1865–1900," Economic History Association, accessed February 4, 2025, https://eh.net/encyclopedia/the-economics-of-american-farm-unrest-1865-1900/.
9 Kenneth Finegold, "From Agrarianism to Adjustment: The Political Origins of New Deal Agricultural Policy," *Politics & Society* 11, no. 1 (1982): 7.
10 Thomas Jefferson, *Notes on the State of Virginia* (Penguin Books, 1999), 170–171.
11 Louis Bernard Schmidt, "Andrew Jackson and the Agrarian West," *Current History* 28, no. 166 (June 1955): 321.
12 Peter Lamborn Wilson, "Grange Appeal," *Capitalism Nature Socialism* 17, no. 3 (2006): 28–29, 31.
13 Jeff Taylor, "Agrarian Politics and the American Tradition," in *Localism and the Mass Age: A Front Porch Republic Manifesto*, eds. Mark T. Mitchell and Jason Peters (Front Porch Republic Books, 2018), 81.
14 Phillips, *This Land*, 36.
15 Donald Worster, *Nature's Economy: A History of Ecological Ideas* (2nd edition) (Cambridge University Press, 1994), 221.
16 Worster, *Nature's Economy*, 221–222.
17 Tim Lehman, *Public Values, Private Lands: Farmland Preservation Policy 1933–1985* (University of North Carolina Press, 1995), 11.
18 Hugh Hammond Bennett, "Soil Erosion a National Menace: Some Aspects of the Wastage Caused by Soil Erosion (Part One)," United States Department of Agriculture, Circular No. 33 (April 1928), 18.
19 Phillips, *This Land*, 52, 57.
20 Finegold, "From Agrarianism to Adjustment," 19.
21 Lehman, *Public Values*, 18.
22 "About Districts," National Association of Conservation Districts, accessed November 4, 2024, https://www.nacdnet.org/about-nacd/about-districts/#:~:text=Across%20the%20United%20States%2C%20nearly,%2C%20water%2C%20forests%20and%20wildlife.
23 Phillips, *This Land*, 109–118.
24 Lehman, *Public Values*, 35.
25 Tim Lehman, "Public Values, Private Lands: Origins and Ironies of Farmland Preservation in Congress," *Agricultural History* 66, no. 2 (1992): 257–272.
26 Aldo Leopold, *A Sand County Almanac: And Sketches Here and There* (Oxford University Press, 1987), quotes from pp. 204, 206, 203, and 204 in sequence.
27 Roderick Frazier Nash, *The Rights of Nature: A History of Environmental Ethics* (University of Wisconsin Press, 1989), 63.
28 Daniil Imhoff, *The Farm Bill: A Citizen's Guide* (Island Press, 2019), 36.
29 Barry J. Barnett, "The U.S. Farm Financial Crisis of the 1980s," *Agricultural History* 74, no. 2 (2000): 366–380.
30 David Goldfield, "The Stages of American Urbanization," *OAH Magazine of History* 5, no. 2 (1990): 26. Also, Cynthia Nickerson, Robert Ebel, Allison Borchers, and Fernando Carriazo, "Major Uses of Land in the United States," United

States Department of Agriculture, 2011, https://ers.usda.gov/sites/default/files/_laserfiche/publications/44625/11159_eib89_2_.pdf?v=20473.

31 Lehman, *Public Values*, 19.
32 Michael Bunce, "Thirty Years of Farmland Preservation in North America: Discourses and Ideologies of a Movement, *Journal of Rural Studies* 11 (1995): 238.
33 "Census of Agriculture, Vol. 1," United States Department of Commerce (1992), 8.
34 USDA, Economic Research Service, https://www.ers.usda.gov/.
35 Pamela Riney-Kehrberg, *When a Dream Dies: Agriculture, Iowa, and the Farm Crisis of the 1980s* (University of Kansas Press, 2022), 134–135.
36 Riney-Kehrberg, *When a Dream Dies*, 135.
37 Riney-Kehrberg, *When a Dream Dies*, 143–144.
38 "Characteristics of Small and Rural School Districts," National Center for Education Statistics, May 1997, https://nces.ed.gov/pubs97/97529.pdf.
39 Robert Wuthnow, *The Left Behind: Decline and Rage in Small-Town America* (Princeton University Press, 2018), 60.

Bibliography

Barnett, Barry J. 2000. "The U.S. Farm Financial Crisis of the 1980s," *Agricultural History* 74, no. 2. doi:10.1215/00021482-74.2.366.

Bennett, Hugh Hammond. 1928. "Soil Erosion a National Menace: Some Aspects of the Wastage Caused by Soil Erosion (Part One)," United States Department of Agriculture, Circular No. 33.

Bunce, Michael. 1995. "Thirty Years of Farmland Preservation in North America: Discourses and Ideologies of a Movement," *Journal of Rural Studies* 11. doi:10.1016/s0743-0167(97)00035-1.

Finegold, Kenneth. 1982. "From Agrarianism to Adjustment: The Political Origins of New Deal Agricultural Policy," *Politics & Society* 11, no. 1. doi:10.1177/003232928201100101.

Goldfield, David Goldfield. 1990. "The Stages of American Urbanization," *OAH Magazine of History* 5, no. 2. doi:10.1093/maghis/5.2.26.

Harley, C. Knick. 1978. "Western Settlement and the Price of Wheat, 1872–1913," *The Journal of Economic History* 38, no. 4. doi:10.1017/s0022050700087131.

Imhoff, Daniel. 2019. *The Farm Bill: A Citizen's Guide*. Island Press.

Jeferson, Thomas. 1999. *Notes on the State of Virginia*. Penguin Books.

Lehman, Tim. 1992. "Public Values, Private Lands: Origins and Ironies of Farmland Preservation in Congress," *Agricultural History* 66, no. 2.

Lehman, Tim. 1995. *Public Values, Private Lands: Farmland Preservation Policy 1933–1985*. University of North Carolina Press.

Leopold, Aldo. 1987. *A Sand County Almanac and Sketches Here and There*. Oxford University Press.

Nash, Roderick Frazier. 1989. *The Rights of Nature: A History of Environmental Ethics*. University of Wisconsin Press.

Pfeffer, Max J. 1983. "Social Origins of Three Systems of Farm Production in the United States," *Rural Sociology* 48, no. 4.

Phillips, Sarah T. Phillips. 2007. *This Land, This Nation: Conservation, Rural America, and the New Deal*. Cambridge University Press.

Rasmussen, Wayne D. 1965. "The Civil War: A Catalyst of Agricultural Revolution," *Agricultural History* 39, no. 4.

Riney-Kehrberg, Pamela. 2022. *When a Dream Dies: Agriculture, Iowa, and the Farm Crisis of the 1980s*. University of Kansas Press.

Stewart, James I. 2008. "The Economics of American Farm Unrest, 1865–1900." In *EH.Net Encyclopedia*, edited by R. Whaples. Economic History Association.

Stoll, Steven. 2003. *Larding the Lean Earth: Soil and Society in Nineteenth-Century America*. Hill & Wang.

Taylor, Jeff. 2018. "Agrarian Politics and the American Tradition." In *Localism and the Mass Age: A Front Porch Republic Manifesto*, edited by Mark T. Mitchell and Jason Peters. Front Porch Republic Books.

Wilson, Peter Lamborn. 2006. "Grange Appeal," *Capitalism Nature Socialism* 17, no. 3. doi:10.1080/10455750600874506.

Worster, Donald. 1994. *Nature's Economy: A History of Ecological Ideas*, 2nd edition. Cambridge University Press.

Wuthnow, Robert. 2018. *The Left Behind: Decline and Rage in Small-Town America*. Princeton University Press.

4 American Farmland Preservation and the Spirit of the Law

To understand the American approach to farmland preservation, we must first understand the country's legal principles regarding private property and private property rights. The eighteenth-century sociologist Montesquieu called these principles the *esprit de lois*, or the "spirit of the law," a term he used to describe the way the law in a place is shaped by its customs, morals, and geography. And to understand the American spirit of the law regarding private property and private property rights, we must understand the influence of John Locke.

Locke was an English philosopher who lived from 1632 to 1704. Widely regarded as one of the Enlightenment's most influential thinkers, his ideas on inalienable natural rights were immensely influential on the American founders and the framers of its Constitution. This influence is most directly observed in Thomas Jefferson's famous opening to the Declaration of Independence: "We hold these truths to be self-evident, that all men are created equal, that they are endowed by their Creator with certain unalienable Rights, that among these are Life, Liberty, and the Pursuit of Happiness."

The famous phrase is a rendering from Locke, who wrote in his work titled *Two Treatises on Government*, "Being all equal and independent, no one ought to harm another in his life, health, liberty, or possessions." Today, many scholars presume that Jefferson deliberately substituted the term "pursuit of happiness" for "possessions" in the Declaration of Independence to avoid an association with slavery implicit in the word "possessions," and so changed the Lockean triad to "pursuit of happiness" as a rhetorical flourish.[1]

Nevertheless, Locke's influence on Jefferson is undeniable, as it is on Jefferson's fellow Virginians George Mason and James Madison. In drafting the Declaration, Jefferson was almost certainly influenced by Mason's "Virginia Declaration of Rights" (adopted on June 12, 1776, less than a month before the Declaration of Independence), which referred to the "enjoyment of life and liberty, with the means of acquiring and possessing property, and pursuing and obtaining happiness and safety," another obvious reference to Locke.[2]

Like previous philosophers such as Hobbes, Locke's concept of natural law expressed the idea that there were certain moral truths that applied to all people, regardless of their time, culture, or government. Locke's view of

DOI: 10.4324/9781003644545-5

natural rights is a universalistic one. Humans can discern this law by using their reason, which allows access to the fundamental truth that each has a right to use and preserve his own body, mind, and life. Importantly for the founding fathers, Locke considered these rights inalienable (or "unalienable" in Jefferson's version), meaning they were God-given, or divinely endowed, and could not be taken away by law or government. In Locke's view, all rights ultimately derive from the right to self-preservation, including the right to property. He argued that, because one owns their body, the labor or production of that body are also theirs by right, as is the land or property where that labor was produced.

As Locke puts it:

> Every man has a *Property* in his own *Person*. This no Body has any rights to but himself. The *Labour* of his Body and the *Work* of his Hands, we may say, are properly his. Whatsoever then he removes out of the State that Nature hath provided, and left it in, he hath mixed his Labour with, and joined to it something that is his own, and thereby makes it his *Property*. [3]

Prior to the founding of America, western rulers were kings and aristocrats, who typically determined the extent of a common citizen's rights or liberties. But Locke asserted that our rights derive from nature, not nobles, and cannot be taken away or infringed on arbitrarily. For Locke, the purpose of government was to secure and protect the natural rights of its people under reasonable terms and circumstances. This created a social contract between government and its people—so long as their natural rights are protected, the people have a responsibility to obey the laws of their government. When these rights are abused or ignored, however, the people have a right to resist or abolish that government, creating a new one in its place.

Clearly, Locke's concept of property rights was not absolute. He believed that natural rights are not conveyed in a moral vacuum. Rather, they presumed a framework of ethical responsibility to others. They also presumed a responsibility toward using property for a productive purpose that somehow benefited the broader community. But these moral distinctions hardly mattered in the rhetorical justifications for American independence. For men like Jefferson and Mason, Locke's concept of inalienable property rights coupled with his theories on the role of government were both convenient justification for the cause of independence and conveniently popular among a people hungry for the ownership of land.

But let us not presume that, outside of the realm of political rhetoric, America's spirit of the law toward Locke's vision of inalienable rights was absolute ... or consistently applied. Indeed, for most American settlers, it was land, not inalienable rights, that they sought in the New World. In Europe, land was the principal source of social status and economic power. But this power and status were out of reach for ordinary people, who were typically tenants of land, not owners. For them, the opportunity to acquire land in America meant not only social status and economic opportunity but also an almost inconceivable degree of personal freedom—an opportunity to achieve some measure of the "hegemonic" liberty enjoyed by the Europe's land-owning class.

American immigration not only offered access to this hegemonic liberty; it also offered an abundance of land, and in some cases, colonial administrations that were eager to provide it to new arrivals. For instance, in South Carolina, where England's royal administration and her proprietors wanted to establish a stable and orderly society in the colony's interior "back-country," ordinary settlers could apply for a "headright," an early version of a homestead. The number of acres in a headright grant was determined by the size of the grantees' family and the grantees' ability to cultivate the lands. Interestingly, this headright system was developed by John Locke, who drafted the "Fundamental Constitution of the Carolinas" while serving as secretary to the Earl of Shaftesbury, one of the original eight "Lords Proprietors" who had been granted title to the huge tract of land in North America that eventually became the Province of Carolina in 1669.[4]

Certainly, not every landowner in America was seeking this liberty, for the American colonies contained more than their fair share of speculators, who acquired land not to produce on it or mix with it his or her "Labour," but as an investment. And many immigrants arrived in America without any hope or promise of liberty at all, such as African slaves. Nevertheless, the status of land ownership as both a pathway to "liberty" and an arbiter of it was well established by the late colonial period.

In his book *The Americans: The Colonial Experience,* author Daniel J. Boorstin describes how the hegemonic rights of land ownership transferred to political power in the House of Burgesses, the governing assembly of colonial Virginia. "It was generally accepted in Virginia in those days that the ruling planters of good family had a prescriptive right to become ruling Burgesses," Boorstin writes. "Land—land to use, to waste, to divide among one's children—was the foundation of all the governing families and the fortunes of Virginia." In Virginia's agrarian economy, this power was necessary for accessing profitable commodity markets for tobacco, timber, and livestock, but also in the maintenance of the social status quo. Unlike in England, where the removal of restrictions on inherited land required court authority, the Virginia Burgesses could pass private acts to do so, facilitating an heirs' ability to use and sell the land for profit. Boorstin counted 125 such acts between 1711 and 1774, the vast majority benefiting fellow Burgess members.[5]

Land as an inalienable right. Land as a pathway to wealth and status. Land as an access mechanism to political power. "For three hundred years American popular culture has strongly linked land ownership with individual freedom," writes legal scholar Lynton Caldwell.

> Because our culture has so strongly linked land to personal freedom and wealth, to the new American way of life, landowners in America have long jealously guarded their rights ... perhaps in no other area have popular attitudes more persistently influenced public law and policy.[6]

But private property rights in America have never been as absolute as they are popularly depicted. Through eminent domain, American governments have always held authority to take private property without the owner's consent, provided the taking is for public use and the owner is fairly compensated. For generations, both private corporations and individuals employed eminent domain to create the infrastructure for Manifest Destiny and the Industrial Age. Private railroads used it to construct rail lines on private land; irrigators to access water and miners to access minerals; and governments for major renovations.[7]

In Colonial times, people were permitted by law to enter another's private property to hunt, forage, and graze their livestock under many circumstances, a common practice that continued well into the nineteenth century.[8] Some vestiges of these communal property rights remain. In Vermont, citizens still have a right to hunt on the private "unenclosed lands" of others, if those lands are not posted "No Trespassing." In North Carolina, like many other states, canoers and kayakers can boat across private land provided the waterway is considered "navigable." And through the Endangered Species Act, the United States government has been largely successful in preventing private landowners from destroying habitats for threatened animal and plant communities. Perhaps most significantly, laws permit courts to stop practices on private land that damage or inhibit the property rights of others, such as the spread of harmful chemicals or industrial activities that threaten the health of neighboring landowners and communities.

For land-use scholar Eric Freyfogle, these exceptions symbolize the malleability of private property concepts in our legal system.

> Private property differs from the other individual rights that we view as critical in the United States. It is not free speech, freedom of religion, or the right to a trial by jury. It is a more social, malleable, and indirect kind of right. In operation, it is less an individual right than a tool society uses to promote overall social good.[9]

Nevertheless, it is true that over the last 250 years, the American spirit of the law toward property rights has become more absolute. In the twentieth century, property was increasingly viewed as a private entitlement existing in a private realm. Evoking Locke's "natural rights" arguments, Freyfogle observed:

> If property rose apart from government, if it was an individual right that arose from the mists of time before lawmakers came along, then the way

to get clear about the rights of landowners was to start with basic principles and axioms of individual liberty and equality, and then to reason one's way logically to a scheme of rights.[10]

The current spirit of American property law is perhaps most prominently represented by the Fifth Amendment of the American Constitution, which states in part: "No person shall be ... deprived of life, liberty, or property, without due process of law; nor shall private property be taken for public use, without just compensation." The last portion of this clause, referring to the taking of private property, is referred to in the American legal system as the "Takings Clause."

But there is almost no historical documentation providing elaboration on what the Framers intended by the phrase "taken for public use." Indeed, over 250 years after the ratification of the United States Constitution, that meaning is still being debated by scholars, lawyers, and courts.[11] There is little else about private property in the Constitution, and as a result of the "Takings Clause's" ambiguity, the interpretation and application of American private property law has fallen largely to local courts in the states and counties.[12]

For guidance then, many of these bodies turned to English and European property laws, brought to America by her European settlers. Indeed, for rich aristocrats such as the English Lord Proprietors, or William Penn, who had been gifted a generous land charter from England's King Charles II to establish a Quaker settlement in what is now Pennsylvania, some form of law for establishing the conveyance and ownership of property was essential to their business plans, for they hoped to gain riches by subdividing and selling those tracts to European settlers.

From this English influence arose much of America's current real estate laws. The American laws were not the same, of course, for English law treated land as a perpetual asset, reflecting a society in which political and social authority was vested in a landed class that sustained itself through long-term property ownership and primogeniture, the right of succession belonging to the first-born child. In contrast, American places tended to view real estate law as a mechanism for espousing republicanism and liberty by offering more participation in land ownership than European ones, and thus, more opportunities for political participation.

But many of the real property concepts from America's spirit of the law were the same as England's, and the "bundle of rights" was one of the most influential among them. The phrase's origins lie in the ancient English medieval tradition of the "livery of seizin." In this ceremony, the grantor—typically an English lord or nobleman—conveyed possession of the land or the right to use the land by separating a branch from a bundle of twigs and handing it over to the tenant. In some accounts, the ceremony included a piece of turf and an oath by the grantor or an English property official that began with, "This turf and twig I give to thee ..." The livery (a medieval word meaning "delivery") of seizin (meaning "possession," especially of land) remained the dominant method of transferring land rights until the early sixteenth century, when recorded deeds became the norm, although the ceremony continued to be legal in England until 1925.

In absence of established legal infrastructure, the practice continued in the American colonies until the late seventeenth century. William Penn performed the livery of seizin ceremony with twig and turf when he first arrived in America, symbolizing his possession of what would become Pennsylvania (in essence he gave the land to himself, although under authority of the English crown).[13]

The concept of a "sole and despotic" relationship between people and property remained popular through William Blackstone's *Commentaries on the Laws of England*, an English legal text widely influential in early America. Blackstone was an eighteenth-century English jurist strongly influenced by Locke. His classical liberal definition of private property framed its sole function as a means to secure freedom and autonomy for individuals; their only obligation was to do no harm to others in the exercise of those rights. But by the end of the nineteenth century, Blackstone's interpretation failed to explain the emerging and increasingly intangible relationships between people and property developing in the United States—more complex property interests such as easements, rights-of-ways, and leases were becoming increasingly common, but also business concepts such as goodwill, trademarks, and corporate shares.

To address these complexities, the American Law Institute (ALI) revived the bundle of rights concept in the 1930s to help explain the legal relationships that had defined property law since the end of manifest destiny. As a metaphor, the bundle of rights signifies that property is a set of legal relationships among people, more than mere "ownership."[14] Legal scholar A.M. Honoré also popularized the "bundle of rights" metaphor in an influential 1960s essay in which he attempted to quantify what he called the "incidents of ownership." These included the right to possess, the right to use, the right to manage, and eight others. Honoré's "incidents" demonstrate the concept of property rights that aggregate to explain property ownership.[15]

In the American farmland preservation movement, the "bundle of rights" remains a convenient metaphor for explaining the legal principles behind conservation easements. "In essence, granting a conservation easement creates a legal partition of the ownership bundle of rights," notes legal scholar Julie Ann Gustanski. "The recipient organization retains the right associated with development. In turn, the landowner continues to hold fee simple title to the property, with an understanding that uses must be in keeping with the terms of the easement."[16]

In practice, conservation easements are rarely expressed so eloquently or fully understood by the landowners who convey them. Nevertheless, their basis in the bundle of rights framework, and the influence of John Locke's "inalienable" property rights in their heritage, make conservation easements a legal tool that honors the American spirt of the law toward property rights even as they expand that spirit to the realm of environmentalism and natural resource protection.

Social Rights and the Public–Private Divide

If the American spirit of the law demands that farmland preservation be viewed through the perspective of private property rights, American culture often frames farmland preservation through a broader concept of rights that includes farmland's role as a natural, economic, and social resource benefitting all strata of American society.

"America's farmland is irreplaceable," states one such framing on the American Farmland Trust website.

> It grows our food. It supports wildlife and biodiversity. It cleans our water, increases resilience to natural disasters like floods and fires, and helps combat climate change. Our food, our water, our environment, our survival—it all depends on American farmland and ranchland.[17]

Unlike Locke's "inalienable" personal rights, this is the rhetoric of "social rights," the idea that societies and communities have a collective right to health, prosperity, and quality of life. When applied to the issue of farmland preservation, this view proposes that land use practice should benefit both private individuals and the broader community. "Whether a land use system evidences some concern over the future consequence of present land use decisions depends heavily on whether rights under the system accrue primarily to individuals or society," observes Caldwell. "Concerns with individual rights focus on the present; concerns with social rights, by contrast, logically extend into the future."[18]

In evoking the strategic, economic, and social importance of farmland preservation, social rights rhetoric has been both implicit and explicit in the movement's messaging of the last hundred years. It is explicit in the words of Franklin Delano Roosevelt, who declared in one of his famous fireside chats that, "The farmers of America want a sound national agricultural policy in which a permanent land use program will have a permanent place,"[19] just as it is in the "No Farms, No Food" motto of the American farmland trust. More implicit is the rhetoric linking farmland preservation to rural character and communities, a common theme of "New Agrarian" literature, but the idea is the same.

The scholar David D. Danbom refers to these efforts to link agricultural preservation to the collective benefit of America society and economics through the term "Rational Agrarianism." This Rational Agrarianism, Danbom writes, stresses the contributions agriculture and rural communities provide to America's overall economic and political health. In a broader and more existential sense, the social rights perspective is also reflected by those who call for more sustainable and resilient land uses, and the policies to support them, in response to the environmental impacts of industrial farming and other land-use concerns. Social rights are even embedded in the transcendental themes of "Romantic Agrarianism," which equates the loss of farmland and societal isolation from the land-environment-food system to a corresponding decline in community, mental health, and overall well-being of American society.[20]

However, Rational Agrarianism is also a conception of rights that relies on a presumption of shared values, positioning farmland preservation, and other agendas of the American environmental and conservation movements, as moral and ethical necessities—at least if you share a value system that prioritizes those movements. Many Americans *do* share these values, but there is a political danger in presuming them, or in presuming a specific characterization of them. And many Americans resent having those presumptions cast upon them, or resent being cast as antagonists in them, particularly when they are considered imposed by an American "elite" class who have nothing to do with production of the food they depict as so morally and ethically important.

In American politics, this "values" conflict is often depicted as the "rural/urban" divide, or the "rural/elite" divide, insinuating that America's urban areas are the domain of its educated and socially privileged class. Such characterizations are not wholly inaccurate. In some ways, the rural/urban divide still resonates at the core of American political identities, its differences challenging our ability to work together on society's "wicked" problems such as food systems and climate change.[21]

But the political scientist Michael J. Sandel encourages us to think of this dichotomy as a distinction between the "right" and the "good." Those who believe in the "right" prioritize a fair framework of legal rights within which individuals and groups can choose their own values. This is essentially a Lockean conception of governance and morals. In contrast, those that favor the "good" demand government policy that prioritizes some value-based conception of goodness, even if the "good" infringes on individual liberties and choices. When those Sandel describes as "contemporary liberals" expound the environmentalism, equity, sustainability, and other values implicit in the concept of social rights, they call on a universal conception of "good" that is not necessarily shared by rural and conservative Americans. Naturally, many rural and conservative Americans respond with the arguments of basic rights and liberties embedded within the American spirit of the law, including its premise of "inalienable" property rights, when these presumptions of the moral "good" are imposed upon them.[22]

Like other issues in American society, the arguments for and against farmland preservation often come down to this distinction between the "right" and the "good"—between a framework of basic rights and liberties and conceptions of good people may choose to pursue within that framework. Such is a primary philosophical argument within both America's farmland preservation movement and her spirit of the law: how to adapt to the fundamental changes in America's society, culture, economy, and environment within the rights framework upon which the country was established; and how to advance some policy of social rights for farmland preservation that doesn't offend the values of rural America. For organizations like the American Farmland Trust, the answer lies in focusing on farmland preservation as a "resource" issue. Americans need food to survive, it argues, and to grow food it needs farmland. But this "resource" argument opens itself up to criticism by those who argue that the American agricultural system is doing just fine, thank you very much, at least in terms of production.

There is no simple answer to this dilemma. Indeed, farmland preservation wouldn't be a "wicked" problem if there were. Freyfogle, for one, believes that the American legal system has always contained a more malleable, social spirit than is typically depicted in the "inalienable rights" arguments of Jefferson and Locke. "The reality, as philosophers have long concluded, is that private property in land really has only one solid moral justification. Private ownership is sound because it is useful to us collectively as a people," he believes.[23] American property rights law must change, he argues, because America itself is changing so rapidly.

The political sociologists Nicholas Jacobs and Daniel Shea, however, advocate a more pragmatic approach, adapting the "universal" values of conservation and sustainability to the unique conditions and personalities of the rural communities where farmland preservation must take place. Rural Americans are different, they argue, "because their politics, attitudes, beliefs, and behaviors are shaped by a sense of place." Don't impose on these rural communities a universal "good" dictated by urban elites, they suggest, but a "good" that reflects the values and needs of these individual places.

We shall explore the policy and practice implications of this rural sense of place later in this book. For now, this chapter closes with the indisputable observation that the spirit of the law dominating American property rights for two hundred years is changing but still essential, and the farmland preservation movement is both an instigator and beneficiary of that change. In many ways, the idea that property rights can and should benefit the public good, the social rights, of our communities is nothing new. But the idea that traditional property rights perspectives can be adapted to protect farmland and farming communities through conservation easements is something relatively new.

But is it enough? Two things are certain: the American "inalienable" property rights faction isn't going anywhere any time soon; and the American spirit of the law changes gradually, over decades and centuries, not years. Therefore, what may be necessary for truly groundbreaking advances in farmland preservation for the rest of the twenty-first century are not necessarily changes in America's spirit of the law but its role in addressing the social problems of America's rural farmers, families, and farming communities.

Notes

1 Carli N. Conklin, "The Origins of the Pursuit of Happiness," *Washington University Jurisprudence Review* 7, no. 2 (2015): 195–261.

2 Joseph J. Ellis, "The Spring of '76: Texts and Contexts." In Joseph J. Ellis, Ed., *What Did the Declaration Declare?* (Bedford/St. Martin's, 1999), 79–94.

3 John Locke, *Two Treatises of Government*, Peter Laslett, Ed. (Cambridge, 1960). Excerpted from Kristin Sharder-Frechette, "Locke and Limits on Ownership," *Journal of the History of Ideas* 54, no. 2 (1993): 201–219. doi:10.2307/2709979.
4 Mary C. Simms Oliphant, "Granting of Land in Colonial South Carolina," *The South Carolina Historical Magazine* 77, no. 3 (1976): 208–212.
5 Daniel J. Boorstin, *The Americans: The Colonial Experience* (Random House, 1958), 118–120.
6 Lynton Keith Caldwell, "Land and the Law: Problems in Legal Philosophy," *University of Illinois Law Review* 319 (1986): 320.
7 Stephen Freyfogle, *On Private Property* (Beacon Press, 2007), xx.
8 Brian Sawers, "Keeping Up with the Joneses: Making Sure Your History Is Just as Wrong as Everyone Else's," *Michigan Law Review First Impressions* 111 (2013): 23.
9 Freyfogle, *On Private Property*, xiv.
10 Freyfogle, *On Private Property*, 52.
11 Maureen E. Brady, "The Damaging Clauses," *Virginia Law Review* 14, no. 3 (May 2018): 347.
12 Brady, "The Damaging Clauses," 343, 349–350.
13 John A. Gose, "Real Estate Conveyances from Livery of Seisin to Electronic Transfer: Real Estate Transactions Enter the Digital/Electronic World," *Real Estate Issues* 33, no. 2 (2008): 59–65.
14 Denise R. Johnson, "Reflections on the Bundle of Rights," *Vermont Law Review* 32 (2007): 249.
15 Denise R. Johnson, "Reflections on the Bundle of Rights," 249. Honoré's other "Incidents" are the right to capital, power to alienate, transmissibility, prohibition of harmful use, liability to execution, and residuality.
16 Julie Ann Gustanski, "Protecting the Land: Conservation Easements, Voluntary Actions, and Private Lands," in *Protecting the Land: Conservation Easements Past, Present, and Future*, eds. Julie Ann Gustanski and Roderick H. Squires (Island Press, 2000), 16.
17 "Why Save Farmland," American Farmland Trust, accessed August 27, 2024, https://farmlandinfo.org/why-save-farmland/.
18 Caldwell, "Land and the Law," 323.
19 Franklin Delano Roosevelt, "September 6, 1936: Fireside Chat on Farmers and Laborers," Miller Center at the University of Virginia, https://millercenter.org/the-presidency/presidential-speeches/september-6-1936-fireside-chat-8-farmers-and-laborers.
20 David B. Danbom, "Romantic Agrarianism in Twentieth-Century America," *Agricultural History* 65, no. 4 (1991).
21 Nicholas F. Jacobs and Daniel Shea, *The Rural Voter: The Politics of Place and the Disuniting of America* (Columbia University Press, 2024).
22 Michael J. Sandel, "The Political Theory of the Procedural Republic," *Revue de Métaphysique et de Moral* 93, no. 1 (1988): 60.
23 Freyfogle, *On Private Property*, xviii, 20.

Bibliography

Boorstin, Daniel J. 1958. *The Americans: The Colonial Experience*. Random House.

Brady, Maureen E. 2018. "The Damaging Clauses," *Virginia Law Review* 14, no. 3.

Caldwell, Lynton K. 1986. "Land and the Law: Problems in Legal Philosophy," *University of Illinois Law Review* 319.

Conklin, Carli N. 2015. "The Origins of the Pursuit of Happiness," *Washington University Jurisprudence Review* 7, no. 2.

Danbom, David B. 1991. "Romantic Agrarianism in Twentieth-Century America," *Agricultural History* 65, no. 4.

Ellis, Joseph J. 1999. "The Spring of '76: Texts and Contexts." In *What Did the Declaration Declare?*, edited by Joseph J. Ellis. Bedford/St. Martin's.

Freyfogle, Stephen. 2007. *On Private Property*. Beacon Press.

Gose, John A. 2008. "Real Estate Conveyances from Livery of Seisin to Electronic Transfer: Real Estate Transactions Enter the Digital/Electronic World," *Real Estate Issues* 33, no. 2.

Gustanski, Julie Ann. 2000. "Protecting the Land: Conservation Easements, Voluntary Actions, and Private Lands." In *Protecting the Land: Conservation Easements Past, Present, and Future*, edited by Julie Ann Gustanski and Roderick H. Squires. Island Press, 2000.

Jacobs, Nicholas F. and Daniel Shea. 2024. *The Rural Voter: The Politics of Place and the Disuniting of America*. Columbia University Press.

Johnson, Denise R. 2007. "Reflections on the Bundle of Rights," *Vermont Law Review* 32.

Locke, John. 1960. *Two Treatises of Government*. Edited by Peter Laslett. Cambridge University Press.

Oliphant, Mary C. Simms. 1976. "Granting of Land in Colonial South Carolina," *The South Carolina Historical Magazine* 77, no. 3.

Sandel, Michael J. 1988. "The Political Theory of the Procedural Republic," *Revue de Métaphysique et de Moral* 93, no. 1.

Sawers, Brian. 2013. "Keeping Up with the Joneses: Making Sure Your History Is Just as Wrong as Everyone Else's," *Michigan Law Review First Impressions* 111.

5 American Agrarianism

Thomas Jefferson outlined his moral, economic, and cultural vision for the American farmer in his 1785 work *Notes on Virginia*: "Those who labor in the earth are the chosen people of God, if ever he had a chosen people, whose breasts he has made his peculiar deposit for substantial and genuine virtue," he wrote.[1]

Today, we acknowledge the moral ambivalences of Jefferson's ideal—the "chosen people of God" he described were White male members of the Virginia gentry, whose "genuine virtue" included the exploitation of slaves, servants, and women. But his characterization of the farmer retains a symbolic power that still resonates in American culture and politics, offering an idealized vision of an American republic founded on the virtue, morals, and industry of her independent farmers.

In *Community and the Politics of Place*, the writer and Montana politician Daniel Kemmis describes how Jefferson's portrait of the American farmer's "genuine virtues" has evolved into a romanticized depiction of civic participation in a republican form of government.

> Jefferson ... believed that a sense of mutual responsibility for one another was a necessary feature of self-government. He believed that the way people made their livings had much to do with the development of such a responsibility, and that farmers developed it most consistently ... In the hard, direct necessities of such agriculture, Jefferson saw the roots of a plain honesty, industry, and perseverance—he saw, in other words, the roots of those "civic virtues" upon which real citizenship depended.[2]

The term "agrarianism" assimilates these symbolic and functional representations of the American farmer. In fact, "agrarianism" is another boundary concept, situated at the intersection of various meanings and interpretations. The *American Heritage Dictionary*, for example, defines agrarianism as "a movement for equitable distribution of land and for agrarian form," while the *Oxford Companion to Modern Poetry* says it is "a movement among men of letters in the American South of the late 1920 and 1930s to defend it traditional agricultural way of life against the industrialization and urbanization that were fast overwhelming it."[3]

DOI: 10.4324/9781003644545-6

It's true that "agrarianism" can refer to a political or a literary movement, among other interpretations, but author Eric T. Freyfogle's definition captures the full breadth of the word's meaning. "Agrarianism, broadly conceived, reaches beyond food production and rural living to include a wide constellation of ideas, loyalties, sentiments, and hopes. It is a temperament and a moral orientation as well as a suite of economic practices, all arising out the insistent truth that people everywhere are part of the land community, just as dependent as other life on the land's fertility and just as shaped by its mysteries and possibilities."[4]

Freyfogle's description encompasses the spiritual, philosophical, and social dimensions of agrarianism, along with its political and literary ones. However, no matter how you define it, agrarianism remains a powerful concept in American politics and policy. If Franklin Delano Roosevelt adopted Jefferson's agrarian symbolism in his campaign speeches and New Deal policies, today's politicians still evoke these agrarian tropes. "Farmers feed America," said former American president Joe Biden at a political event in 2022, "you're literally the backbone of our country." Not to be outdone, President Donald Trump repeated this sentiment during the Covid-19 crisis of his first administration by

> expressing our profound gratitude … to the farmers and producers across the country, who have kept our nation fed and nourished as we have battled this invisible enemy … You remind us once again that the American farmer is the backbone of our country.[5]

On the symbolic power of the American farmer, at least, Democrats and Republicans can agree, reminding us that little has changed since the age of Jefferson in agrarian political rhetoric. However, the scale, scope, and spirt of American agriculture *has* changed dramatically, leading to a foreboding sense of dehumanization, economic decline, and spiritual disintegration in today's rural farming communities.

The writer Lauren Soth attributes the twentieth century's crisis in American agrarianism to three major factors. The first was the rise in technology. As technology allowed farms to consolidate and enlarge, the need for farm labor declined dramatically, creating a historic movement of people from rural to urban areas. Second was the rising income gap between large-scale, industrialized farmers and those who clung to more traditional modes of agriculture. Third was the transition away from the general farming that dominated most of the eighteenth, nineteenth, and early twentieth century to an age of specialization and commodities of scale in agricultural production. As Soth notes, up until the 1940s, as much as 75 percent of America's farms were general farms whose operations had common interests. But today there are more large-scale, specialized farming operations, each with its own economic and political interests.[6]

New awareness about the health and environmental impacts of industrialized farming can now be added to Soth's analysis. Sadly, the chemicals and pesticides necessary for farming's industrial scale frequently wind up in our air and water supply, increasing rates for cancer and other deadly diseases. In the large-scale livestock feeding and processing operations known as CAFOs (an acronym for concentrated animal feed operations), ammonia from animal waste and heavily fertilized fields create airborne particulates that can cause heart and circulatory disease when breathed through human lungs.[7] The antibiotics necessary to maintain animal health in these crowded conditions create pathogen resistance, while overflow from CAFO waste lagoons has become increasingly common in severe weather events, flooding surrounding areas with dangerous toxins. Meanwhile, CAFO manure spread on crops and fields as fertilizer overwhelms the soil, sending excess nutrients into the water supply, damaging aquatic communities (and human health).

Now we can add the dangers of "forever chemicals" to these environmental threats. Treated human waste tainted with forever chemicals is now commonly spread as fertilizer on American farm fields. When livestock graze the fields, these dangerous chemicals become embedded in the food system, creating a deadly cycle of forever chemical distribution between humans, ecosystems, and livestock.

Sadly, the list of health threats to both humans and animals posed by industrial farming goes on, and add to these threats industrial agriculture's impacts on global warming, thanks in part to increased emissions of methane and carbon dioxide from livestock. One study released by the Food and Agriculture Organization of the United Nations estimates that livestock produce 11.1 percent of global greenhouse emissions, while other peer-reviewed studies place the figure as high as 19.6 percent.[8]

Health threats aside, the practices of industrialized agriculture also raise broader ethical and moral questions about the treatment of animals and farm laborers. Many claim the living conditions endured by animals in large-scale CAFO operations are simply immoral. And the laborers in these operations are subject to many of the same health and environmental dangers as the animals, with little government oversight or protection.[9] These farm laborers are also typically underpaid when compared to other labor sectors. According to the Economic Policy Institute, U.S. farmworkers were paid an average of $14.62 per hour as of 2020, less than 60 percent of what comparable workers outside of agriculture made the same year.[10] Not surprisingly, these farm workers are among the most economically disadvantaged and socially vulnerable groups in the United States, with few legal or labor protections, dangerous working conditions, unprotected exposure to toxins, threats to security and family from immigration enforcements, and lack of safe and affordable housing.[11]

Minority and women farm owners and workers also suffer in industrialized farming systems. As farm operations have consolidated, Black farm ownership in America has declined at twice the rate of White farm ownership in the past half century. And a recent report by the American Farmland Trust found that women-led farms receive less government financial and technical support and make less money than their male-led counterparts.[12]

Rise of the "New Agrarianism"

In response to the ethical, health, and safety challenges of "big" agriculture, its dehumanizing scale and harmful environmental practices, a "New Agrarianism" has arisen, focusing on a moral orientation toward modern farming practices, incorporating concepts of equity, restoration, conservation, and reduced resource consumption into America's twenty-first century agricultural practice and policy. Organic and regenerative farming are examples of New Agrarianism, as are efforts to connect urban residents to agrarian values through urban gardens, farm "incubators," and farmer's markets. Programs that enhance access to land and provide training and resources to new and minority farmers are also in the New Agrarian tradition, along with farm-to-school programs that attempt to provide locally grown options for school cafeterias. In its focus on conserving natural resources, protecting small-scale farming, and helping family's retain land, farmland preservation also falls into the New Agrarian ethic.

Writing a generation ago, David Danbom called this movement "Romantic Agrarianism." Romantic Agrarians, Danborn argues, are similar to "Rational Agrarians," who operate in the tradition of Thomas Jefferson and Franklin Delano Roosevelt and stress the tangible contributions agriculture and rural people make to a nation's economic and political well-being. But whether called "New" or "Romantic," this concept incorporates the influences of American writers and thinkers such as Henry David Thoreau, who considered contact with nature in general and the land in particular as the primary benefit of an agrarian lifestyle, as well as the social influences of the Grange, which sought to enhance rural community through fellowship and the fraternal bonds of Masonic-style ritual.

Danbom attributes the rise of Romantic Agrarianism to two factors. The first was as a vehicle for criticizing the capitalistic and technologically oriented practices of big agriculture and the dehumanizing elements of urban life. The second was its embodiment of a particularly strong and alluring theme in American culture, the idea of the "Adamic Myth," or the belief that the individual can recapture a lost innocence by somehow returning to a more pure and innocent lifestyle.[13]

Indeed, in its appeals to our communal human nature, both as farmers and the broader community of the entire food system, New Agrarianism carries on in the socio-cultural tradition of the early Romantic Agrarians, with a new awareness about modern farming's environmental impacts and the rapidly changing nature—some would say decline—of American society and social practices. These social dimensions of New Agrarianism also introduce elements of social capital theory into agricultural policy—modern industrial agriculture not only threatens the independent farmer, it argues, but also the social fabric of farming communities. However, a stark difference between Danbom's "Romantic Agrarians" and Freyfogle's "New Agrarians" is a concern over the environmental impacts of modern farming practices, and an ethical orientation toward the farming ecosystem, its natural organisms and communities, reflected in an orientation toward social-ecological thinking, an understanding of humans and nature as one, co-evolving system. As researcher Hannah Whitman and her co-authors write, the New Agrarianism "builds from the roots of land-based and agrarian citizenship, whereby relationships between people (as both producers and consumers of food) and nature are characterized by a socio-ecological metabolism that is mediated by, but not limited to, access to land and political power."[14]

Aldo Leopold's vision for a social-ecological ethic in farming and land management was one of the major influences on the New Agrarianism movement. Humans, Leopold argued in *A Sand County Almanac*, must view themselves not simply as beneficiaries of the ecological system in which farming occurs, but also active partners. But the exploitive nature of modern agriculture leads to a "progressive and mutual deterioration, not only of the plants and soils, but of the animal community subsisting thereon." What is necessary to address this exploitive nature system is a land ethic, Leopold argues.

> The land ethic simply enlarges the boundaries of the community to include, soils, waters, plants, and animals, or collectively: the land … In short, a land ethic changes the role of Homo sapiens from conqueror of the land community to plain member and citizen of it.[15]

The New Agrarian farmer, novelist, poet, and essayist Wendell Berry is widely considered to be Leopold's most prolific disciple, if not his direct successor. For example, consider Berry's definition of the social-ecological community in comparison to the one of Leopold provided above:

> By community, I mean the commonwealth and common interests, commonly understood of people living in a place, and wishing to continue to do so. To put it another way, community is a locally understood interdependence of local people, local culture, local economy, and local nature.[16]

No surprise, then, that Wendell is one of America's most powerful critics of modern industrial agriculture and its relationship to the land.

> We have given up the understanding … that we and our country create one another, depend on one another, are literally part of one another … all who are living as neighbors here, human and plant and animal, are part of one another.[17]

Agrarian Discourse: Rational vs. Romantic

In *Community and the Politics of Place,* Daniel Kemmis describes the opposing influences of community and what he calls "unrestrained individualism" in public discourse. To frame this conflict, Kemmis describes "the first language of individualism" as the discourse people use to criticize policy or government initiatives in terms of how they and their immediate families are impacted. This is the language in which people

> readily and regularly oppose each other at public hearings, avidly pursuing our own interests and protecting our own rights with no sense of responsibility to hear or respond to the legitimate interests of those on the "other side" or discover common ground.[18]

Kemmis contrasts the language of individualism with what he calls "the second language of commitment." This is the language in which people "come into relationship with other people in a very particular way,"[19] recognizing the ethical dimensions of common interests and the power of collective action.

However, Kemmis cautions, because the first language of individualism is inherent in human nature, the second language of commitment only comes through the "practice of cooperation." From a social capital theory perspective, the second language of commitment can be thought of as the language of the daily interactions and shared experiences that form social bonds, or what will be described as "bonding" social capital in a subsequent chapter. It is the language we use (mostly) in church committees and nonprofit boards, and generally in community processes oriented around collective goals that take the time to recruit and develop citizen stakeholder participants and establish ground rules for cooperative dialog.

But Kemmis reminds us that this cooperative discourse is contingent on the physical dimensions of place, "the actual, specific places within which those practices and that cooperation take place," and warns us that no real collective governance is possible except among people who are of the same place, sharing the same perspectives, intent on solving the same problems. "If there are not habituated patterns of work, play, grieving, and celebration designed to enable people to live well in a place, then those people will have at best a limited capacity for being public with one another," Kemmis warns.[20]

In American policy and political life, the language or discourse of agrarianism is often employed in a broad sense, adopted by politicians who speak of the idealized, romanticized role of the American farmer in pursuit of some political objective. It is also frequently the language of the Rational Agrarians, who talk about farmland preservation in terms of its threat to our national food system. But Leopold's "land ethic" and Berry's "locally understood interdependence" of community speak the second language of commitment, the language of relationship between people and of people within a specific place. It is not an instrumental discourse based on a desired outcome, nor (at its best) the discourse of "right" versus "good," but one grounded in the building of shared values and relationships, the language of commitment to a shared place.

Although New Agrarianism has arisen largely as a criticism against the excesses and abuses of industrial agriculture, it also serves as an expression of Danbom's "Adamic Myth," the quest for a return to a more thoughtful, considerate model of agriculture, and the moral and ethical benefits this quest entails. If, as Danbom believes, "the pursuit of commercial goals at the expense of an emotionally satisfying life has damaged both the family and the community,"[21] then Romantic Agrarians believe that a more holistic and sustainable brand of agriculture can provide antidote to the spiritual degradation of this pursuit.

These themes of community, conservation, and the Adamic Myth can be described as part of New Agrarianism's symbolic capital, a concept derived from Pierre Bourdieu's capital theory. Bourdieu employed the concept of capital as a theoretical framework for the capacity of groups and individuals to impact situations and outcomes. Symbolic capital was Bourdieu's fourth type of capital, along with economic, cultural, and social capital. Each could be accumulated and used by individuals or groups in socially advantageous ways. However, Bourdieu believed that every form of capital tends to function as symbolic capital when it is recognized as legitimate by others. This makes symbolic capital a type of "metacapital" that enables those who exercise it to shape perceptions of the social world.[22]

Over time, Bourdieu defined symbolic capital in different ways. Writing in 1987, he described it as

> nothing but economic or cultural capital as soon as they are known and recognized, when they are known according to the perception categories they impose, the symbolic strength relations tend to reproduce and reinforce the strength relations which constitute the structure of social space.[23]

Translated from French, Bourdieu's writing can often seem puzzling and unnecessarily abstract, even to sociologists. In simpler terms, Bourdieu's concept of symbolic capital describes the symbolic characteristics used in class differentiations, dominations, and reproductions. In his first book, *Sociologie de l' Algérie*, published in 1958, Bourdieu developed his theories on symbolic capital through an ethnographic analysis of Algeria's Arab-speaking tribes. He theorized that the prestige associated with each tribe was perpetuated through stories of their past accomplishments and origins. Over time, these narratives became attached to the tribe's identity, conveying its authority and societal reputation.

For Bourdieu, this narrative or symbolic prestige is a variant of social magic, which he also called "social alchemy," the way that narrative and symbols in social relationships can create outcomes that seem spontaneous, random, or easy, but are actually the product of their symbolic associations.[24] As Bourdieu developed his theories of symbolic capital, he began to believe it was most prevalent in traditional societies, where narrative symbolism was used to maintain familial and personal "honor." Jefferson's famous description of the American farmer is an example of this symbolism, for it argues that the American farmer is both more virtuous than other Americans and more essential to its virtuous nature. This implies a generally positive connotation for symbolic capital, but Bourdieu also believed symbolic capital could assume negative characteristics. He wrote:

> As nobody is entirely and uniformly characterized by a positive or negative symbolic capital, the amount of symbolic capital of an individual or group can be described as a social combination of attractive and repulsive dimensions of this individual, group, etc.[25]

Bourdieu primarily used capital theory to explain the power relations between groups in society. Indeed, the power dimensions of agrarian symbolic capital are evident in politicians like Trump and Biden, who evoke the symbols of American farming to pass legislation or curry favor with certain voter segments. But symbolic capital can also be used in a broader sense, as part of an advocacy for framing the meanings other members of a society ascribe to social and environmental issues.[26]

Place also plays a role in Bourdieu's symbolic capital theories. Towns, regions, states, and countries are not only physical locations, but are also composed of symbolic characteristics related to history, culture, values, and identities. Here in North Carolina, my home state, to say you are from Asheville conveys a symbolism of art, culture, and a passion for the outdoors. An affiliation with the city of Charlotte, in contrast, conveys a symbolism of business and urban growth (although it also has plenty of fine culture and arts). But place-based iterations of symbolic capital can also have negative connotations. If Asheville and Charlotte are attractive places to visit and work, there are other parts of North Carolina where the decline of the tobacco, textile, and furniture industries has created symbolic associations with economic downturn and cultural desolation. But this type of negative symbolism is not necessarily a death sentence for place. Durham, North Carolina, is a good example of a city that was once viewed symbolically as economically depressed and culturally ravaged yet has undergone a renaissance over the last thirty years, making it one of the most desirable locations in the state to work and live. Of course, Durham benefits from its inclusion in the economically vibrant Triangle region and from being the home of Duke University. The point is that the symbolic power of place can be a powerful determinant of its future, but such a future is not set in stone, and symbolic power can be shifted from negative to positive contexts given the right conditions and an effective narrative.

Farmland Preservation as Symbolic Capital

Thomas Jefferson may have been one of the first to recognize the social magic of the American farm, or at least to write about it so articulately, but he was far from the last, and there are echoes of Jefferson's paean to the "cultivators of the earth" in much of the agrarian symbolism, "New" or otherwise, surrounding farmland preservation. When organizations like the American Farmland Trust proclaim, "America's farmland is irreplaceable ... Our food, our water, our environment, our survival—it all depends on American farmland and ranchland,"[27] they may lack Jefferson's literary eloquence, but they are evoking the same rhetorical tradition.

"To Jefferson, agriculture was not primarily a source of wealth, but of human virtues and traits most congenial to popular self government. It had a sociological rather than an economic value," writes the Jefferson scholar A. Whitney Griswold.[28] Certainly Aldo Leopold, Wendell Berry, and other voices of the New Agrarian movement acknowledge this same value, even if they depict it as a social-ecological one, acknowledging perspectives on the fragile relationship between humans and the environment that didn't exist in Jefferson's time.

Clearly, then, farmland preservation functions in a variety of symbolic ways, encompassing different dimensions of America's "agrarian" traditions. In its evocation of Leopold's "land ethic," championed by Berry, Freyfogle, and others, it is framed as an initiative of the New Agrarianism, or Romantic Agrarianism. As a symbol of disintegrating rural communities and economies, it becomes a call for more political and policy support of rural infrastructure and resources. And as a call for protection of the American agriculture system and its resources, it is a symbol of Rational Agrarianism, depicting America's agriculture system as a vital social, economic, and strategic resource.

But if Jefferson's championing of the American husbandman's "substantial and genuine virtue" appeals to something quintessential about the American character, he used this symbolism for his own political purposes, promoting his belief in a limited, decentralized American government. New Agrarians and farmland preservationists have their own political agendas, using farmland as symbolic capital for a more sustainable and ethical brand of farming, one that perpetuates Leopold's land ethic and preserves the environmental benefits of healthy farmland. Community also becomes a symbolic tool of this political agrarianism, for how can farming communities persevere or even thrive without the land on which to farm?

The multifaceted appeal of farmland preservation is a substantial part of its magic or alchemy, imbuing its policy with a bipartisan flavor that is an important component of its political appeal. Whether you call it Rational Agrarianism, New Agrarianism, or Romantic Agrarianism, the policy of farmland preservation is about the struggle for influence and resources, framed within the multidimensional symbolism of the American farm. This is not a value judgment, simply an acknowledgment of the powerful role of symbolic capital in farmland preservation policy. As Bourdieu observed over sixty years ago, we all use symbolic capital for the acquisition of prestige, power, and influence. The only difference is the stories we tell.

Notes

1 Thomas Jefferson, *Notes on the State of Virginia* (Penguin Books, 1999), 170–171.
2 Daniel Kemmis, *Community and the Politics of Place* (University of Oklahoma Press, 1990), 21.
3 *The American Heritage Dictionary of the English Language* (5th edition) (Houghton Mifflin Harcourt Publishing Company, 2011), 34. Also, N. Everett, "Agrarianism." In I. Hamilton and J. Noel-Tod, Eds., *The Oxford Companion to Modern Poetry* (2nd edition) (Oxford University Press, 2014).
4 Eric T. Freyfogle, Ed., *The New Agrarianism: Land, Culture, and the Community of Life* (Island Press/Shearwater, 2001), xiii.
5 Donald Trump, "Remarks by President Trump on Supporting our Nation's Farmers, Ranchers, and Food Supply Chain," May 19, 2020, https://trumpwhitehouse.archives.gov/briefings-statements/remarks-president-trump-supporting-nations-farmers-ranchers-food-supply-chain/.
6 Lauren Soth, "The End of Agrarianism: Fission of the Political Economy of Agriculture," *American Journal of Agricultural Economics* 52, no. 5 (Dec. 1970): 665.
7 "A Major Source of Air Pollution: Farms," The Earth Institute at Columbia Climate School, May 16, 2016, https://www.earth.columbia.edu/articles/view/3281.
8 "Pathways Toward Lower Emissions: A Global Assessment of the Greenhouse Gas Emissions and Mitigation Options from Livestock Agrifood Systems," Food and Agricultural Organization of the United Nations, 2023, https://openknowledge.fao.org/server/api/core/bitstreams/a06a30d3-6e9d-4e9c-b4b7-29a6cc307208/content. Also, "Livestock Don't Contribute 14.5% of Global Greenhouse Gas Emissions," The Breakthrough Institute, March 20, 2023, https://thebreakthrough.org/issues/food-agriculture-environment/livestock-dont-contribute-14-5-of-global-greenhouse-gas-emissions#:~:text=In%20short%2C%20livestock%20production%20appears,land%2C%20or%20land%2Duse%20change.
9 Frederico Castillo et al., "Environmental Health Threats to Latino Migrant Farmworkers," *Annual Review of Public Health* 42, no. 1 (2021): 257–276.
10 Daniel Costa, "The Farmworkers Wage Gap Continued in 2020," Economic Policy Institute, July 20, 2021, https://www.epi.org/blog/the-farmworker-wage-gap-continued in 2020 farmworkers-and-h-2a-workers-earned-very-low-wages-during-the-pandemic-even-compared-with-other-low-wage-workers/.
11 Megan Horst and Amy Marion, "Racial, Ethnic, and Gender Inequities in Farmland Ownership and Farming in the U.S.," *Agriculture and Human Values* 36 (2019): 4.
12 Caitlin Joseph, Gabrielle Roesch-McNally, and Ariel Looser, "The State of Gender Equity in U.S. Agriculture," American Farmland Trust, 2024, https://farmlandinfo.org/wp-content/uploads/sites/2/2023/12/AFT-gender-equity-report.pdf.
13 David B. Danbom, "Romantic Agrarianism in Twentieth-Century America," *Agricultural History* 65, no. 4 (1991): 1.
14 Hannah Wittman, Jessica Dennis, and Heather Pritchard, "Beyond the Market? New Agrarianism and Cooperative Farmland Access in North America," *Journal of Rural Studies* 53 (2017): 304.
15 Aldo Leopold, *A Sand County Almanac: And Sketches Here and There* (Oxford University Press, 1987), quotes from pp. 203, 207, 210, and 204 in sequence.

16 Wendell Berry, *Sex, Economy, Freedom, and Community* (Pantheon Books, 1993), 119–120.
17 Wendell Berry, *The Unsettling of America: Culture and Agriculture* (Sierra Club, 1973), 23.
18 Kemmis, *Community and the Politics of Place*, 73.
19 Kemmis, *Community and the Politics of Place*, 76.
20 Kemmis, *Community and the Politics of Place*, 79–80.
21 Danbom, "Romantic Agrarianism," 2, 10.
22 Guy Feldman, Roni Strier, and Hillel Schmid, "The Performative Magic of Advocacy Organizations," *British Journal of Social Work* 46, no. 6 (2016).
23 Pierre Bourdieu, *Choses Dites* (Minuet, 1987). Quote extracted from F. Lebaron, "Symbolic Capital." In A.C. Michalos, Ed., *Encyclopedia of Quality of Life and Well-Being Research* (Springer, 2014), doi:10.1007/978-94-007-0753-5_2961.
24 Steven Threadgold, "Illusio, Social Gravity and Social Magic: Purpose Motivation and Aspiration," in *Bourdieu and Affect: Towards and Affective Affinities* (Bristol, 2020).
25 F. Lebaron, "Symbolic Capital." In A.C. Michalos, Ed., *Encyclopedia of Quality of Life and Well-Being Research* (Online edition) (Springer, 2014), doi:10.1007/978-94-007-0753-5_2961.
26 Feldman et al., "The Performative Magic of Advocacy Organizations."
27 "Why Save Farmland," American Farmland Trust Farmland Information Center, accessed December 19, 2024, https://farmlandinfo.org/why-save-farmland/.
28 A. Whitney Griswold, "The Agrarian Democracy of Thomas Jefferson," *The American Political Science Review* 40, no. 4 (Aug. 1946): 667.

Bibliography

Berry, Wendell. 1973. *The Unsettling of America: Culture and Agriculture*. Sierra Club.

Berry, Wendell. 1993. *Sex, Economy, Freedom, and Community*. Pantheon Books.

Bourdieu, Pierre. 1987. *Choses Dites*. Minuet.

Castillo, Frederico, 2021. "Environmental Health Threats to Latino Migrant Farmworkers," *Annual Review of Public Health* 42, no. 1. doi:10.1146/annurev-publhealth-012420-105014.

Danbom, David B. 1991. "Romantic Agrarianism in Twentieth-Century America," *Agricultural History* 65, no. 4.

Feldman, Guy, Roni Strier, and Hillel Schmid. 2016. "The Performative Magic of Advocacy Organizations," *British Journal of Social Work* 46. doi:10.1093/bjsw/bcv088.

Freyfogle, Eric T. Ed. 2001. *The New Agrarianism: Land, Culture, and the Community of Life*. Island Press/Shearwater.

Griswold, A. Whitney. 1946. "The Agrarian Democracy of Thomas Jefferson," *The American Political Science Review* 40, no. 4.

Horst, Megan, and Amy Marion. 2019. "Racial, Ethnic, and Gender Inequities in Farmland Ownership and Farming in the U.S.," *Agriculture and Human Values* 36. doi:10.1007/s10460-018-9883-3.

Jefferson, Thomas. 1999. *Notes on the State of Virginia*. Penguin Books.

Joseph, Caitlin, Gabrielle Roesch-McNally, and Ariel Looser. 2024. *The State of Gender Equity in U.S. Agriculture*. American Farmland Trust.

Kemmis, Daniel. 1990. *Community and the Politics of Place*. University of Oklahoma Press.

Leopold, Aldo. 1987. *A Sand County Almanac: And Sketches Here and There.* Oxford University Press.
Soth, Lauren. 1970. "The End of Agrarianism: Fission of the Political Economy of Agriculture," *American Journal of Agricultural Economics* 52, no. 5.
Threadgold, Steven. 2020. "Illusio, Social Gravity and Social Magic: Purpose Motivation and Aspiration." In *Bourdieu and Affect: Towards and Affective Affinities*. Bristol.
Wittman, Hannah, Jessica Dennis, and Heather Pritchard. 2017. "Beyond the Market? New Agrarianism and Cooperative Farmland Access in North America," *Journal of Rural Studies* 53. doi:10.1016/j.jrurstud.2017.03.007.

Part Two

American Farmland Preservation Practices and Policies Today

6 American Farmland Preservation Today

In 2022, the American Farmland Trust (AFT) released the "Farms Under Threat 2040" report, a comprehensive, quantitative study of farmland loss in America. "The earth has undergone unprecedented urbanization in recent decades and the global development of land consumes some of our planet's most productive agricultural lands," the authors wrote in a peer-reviewed summary of the report's findings. "Such losses of agricultural lands already pose a substantial threat to local and regional food security, especially where rapid conversion of croplands is occurring."[1]

The analysis indicated that, between 2001 and 2016, America lost or compromised approximately 2,000 acres of farmland and ranchland every day. However, its emphasis was on a future-looking trend analysis indicating that if urban development continues at the same pace, America will lose an additional 18 million acres of agricultural land by the year 2040, an area the size of South Carolina. And if development increased under a "Runaway Sprawl" scenario, as much as 24.4 million acres might be lost, an area roughly the size of Indiana.[2]

The report's perspective is squarely within the neo-Malthusian rhetoric of the Rational Agrarian, focusing on farmland conversion as a resource crisis. Clearly, it is written to alarm, even if the decline in farmland has resulted in little loss in overall net production of United States agricultural commodities—according to the United States Department of Agriculture (USDA), America produced 49.1 million metric tons of wheat in 2023/2024, just slightly below its ten-year average of 51.37 million metric tons; corn production for the same period was 389.67 million metric tons, above its ten-year average of 364.66 million metric tons.[3] But just because there's not an obvious relationship between farmland loss and commodity food supply doesn't mean such Malthusian warnings aren't effective. Indeed, AFT's "No Farms, No Food" slogan has always been rooted squarely in the rational agrarian perspective, even if its objectives now encompass broader social and environmental perspectives.

A deeper reading of the "Farms Under Threat 2040" report tells a more nuanced story, depicting runaway farmland loss in rapidly urbanizing parts of the county such as the South and Northeast, as well as around isolated

DOI: 10.4324/9781003644545-8

metropolitan areas such as Denver and Dallas, but with relatively minor impacts in "breadbasket" states like Iowa and Kansas where much of America's agricultural commodities are still grown. But states like Iowa and Kansas are still suffering. Those were the states hardest hit by the farm crisis of the 1980s and are American places where the urban/rural divide has grown even more pronounced over the last forty-five years, contributing to American politics' increasingly bitter partisan divides. "The moral outrage of rural America is a mixture of fear and anger," observes the rural sociologist Robert Whutnow.

> The fear is that small-town ways of life are disappearing. The anger is that they are under siege ... It stems from the fact that the social expectations, relationships, and obligations that constitute the moral communities they take for granted and in which they live are year by year being fundamentally fractured.[4]

In their study of rural voters, the authors Nicholas Jacobs and Daniel Shea come to similar conclusions, arguing that America's rural/urban divide is the core issue influencing American politics today. The difference between the rural and urban voter, they surmise, is one of grievance. "This is the image of real America versus everyone else—a rural America on the ropes, distrusting of government officials, resentful of urban residents' decision-making power, and aggrieved by a sense that they are not getting what they deserve."[5]

Rural sociologists like Whutnow, Jacobs, and Shea argue that the changing nature of farming—the industrialization, dynamic technologies, and economies of scale threatening small- and mid-scale farms—is only a contributing factor to this aggrievement, not its primary source. Other factors, such as the rise of digital culture and the loss of rural institutions like churches, schools, and family businesses, play an even larger role. But the loss of family farmland is a powerful symbol of this change, making farmland conversion both a significant outcome and symptom of other forms of rural social decline. This empowers the farmland preservation movement with unique social magic, as both a symbol of rural restoration and reconciliation between this rural/urban divide.

Conservation Easements and Land Trusts

In January 1981, at the dawn of the 1980s farm crisis, the USDA and the Council on Environmental Quality published the National Agricultural Lands Study (NALS). It estimated that American farmland was being converted to nonagricultural use at a rate of 3 million acres per year, a conversion the study deemed "irreversible." Although the report had been established as a bipartisan effort to study the farmland conversion issue, the report's publication coincided with the beginning of the Ronald Reagan administration, whose leaders viewed it as part of the liberal environmental agenda of the preceding Jimmy Carter administration.

The NALS' conclusions inspired passage of the federal Farmland Protection Policy Act (FPPA) later that year, the first legislation that recognized farmland protection as a federal priority. But the Reagan administration lacked the visionary land-use reformers and progressive agrarians of both the New Deal and the latter "New Agrarian" movement, and the FPPA's provisions relied heavily on the voluntary and discretionary practices of the same "New Conservation" ideologies that had failed to achieve significant farmland protection measures over the previous fifty years.[6]

One reason the federal government and most states generally veer toward economic incentives and "right-to-farm" approaches over comprehensive land-use reform measures is the increasing popularity of conservation easements. "A conservation easement is a legal agreement between a landowner and an eligible organization that restricts future activities on the land to protect its conservation values," reads the Land Trust Alliance's *Conservation Easement Handbook*, a definitive source for land conservation practitioners.[7] These preservation tools have been around in different iterations since the late 1800s, but their use escalated widely in 1981 after the National Conference of Commissioners on Uniform State Laws (now the Uniform Law Commission) promulgated the Uniform Conservation Easement Act (UCEA), proposing uniform standards for conservation easement law across states. The act not only recognized the growing practice of putting voluntary land-use restrictions on land; it also inspired new states to adopt these practices. Today all fifty states have statutes enabling conservation easements, with the UCEA considered the dominant source for state statutory conservation easement law.[8]

More about the mechanics and practices of conservation easements will be discussed in the proceeding chapter. For now, we focus on their historical development, tracing their origins to Massachusetts in the late 1880s and 1890s. There, a talented young landscape architect named Charles Eliot proposed the formation of an association to create and manage "reservations" where the Boston citizens could escape the burdens of rapid urbanization to experience nature's restorative powers. Eliot's advocacy led to the 1891 creation of the Trustees of the Reservation, considered to be the first private land conservation organization in the United States. Eliot's later business partner, the renowned landscape architect Frederick Law Olmsted, played an influential role creating the prototype for conservation easements when he crafted legal agreements with private landowners to protect the parkways he designed in and around Boston.[9]

Inspired by the Trustees of the Reservation, other early land conservancies began to spring up across the country. In New Hampshire, the Society for the Protection of New Hampshire Forests and the Squam Lakes Association were formed in the first decade of the 1900s. In California, the Save-the-Redwoods League was formed in 1917 to protect the state's northern redwood groves, while the Highlands Improvement Society was started in 1905 to preserve the natural beauty of Highlands, North Carolina.

However, these forerunner organizations mostly focused on acquiring or receiving donations of smaller, pristine tracts, typically run as public nature preserves, not holding in trust the development rights of private property belonging to others. It was the federal government that continued to develop and refine the idea of conservation easements first conceptualized by Olmsted. During the 1930s, the U.S. Fish and Wildlife Service obtained hundreds of conservation easements in North Dakota, South Dakota, and Minnesota for the preservation of refuge areas through the Land Utilization Program. That same decade, the National Park Service began acquiring conservation easements in Virginia, North Carolina, Alabama, Mississippi, and Tennessee to protect scenic views along the Blue Ridge and Natchez Trace parkways.[10]

Information is not clear regarding the first nonprofit land trust to "hold" a conservation easement on a privately owned piece of property in the United States. However, the first national land trust was formed in 1946. Originally known as the Ecologists Union, this organization would change its name to the Nature Conservancy in 1950, eventually becoming the most prominent land trust (and much more) in the world. Meanwhile, twenty-eight land trusts were formed between the years 1945 and 1960.[11]

In 1959, California became the first state to pass an enabling statute for conservation easements, and New York followed suit in 1960. In 1964, the Internal Revenue Service (IRS) authorized the first federal charitable income tax deduction for donated conservation easements. A year later, in 1965, Congress passed the Federal Highway Beautification Act, which required states to use 3 percent of their federally allocated funds for landscaping or scenic enhancement, inspiring even more states to pass statutes enabling conservation easements.

Following a number of temporary provisions, the IRS made the conservation easement income tax deduction provision a permanent part of the Internal Revenue Code in 1980. Pursuant to Internal Revenue Code 170(h), a landowner donating a qualified perpetual conservation easement to a government entity or charitable organization is eligible for a federal charitable tax deduction equal to the value of the easement, determined by a qualified appraisal. In order for the landowner to receive tax benefits, the easement must be donated in "perpetuity."

Passage of the UCEA and IRS Code 170(h) helped standardized conservation easement practice and regulations, prompting an explosion of nonprofit land trusts across America. In 1980, there were less than a hundred land trusts operating in the country; by the end of the millennium, that number had risen to over 1,200. And according to Land Trust Alliance (LTA), a national umbrella organization for America's land trust community, there are almost 1,300 land trusts in the United States today, of which just under 1,000 are LTA members.[12]

Also growing were the different types of land targeted by land trusts for conservation easement protection. If the original conservation easements were focused on parks and pristine views, agrarians soon adapted the tool for agricultural preservation, with impressive results. According to the American Farmland Trust's "Protected Agricultural Lands Database," nearly 7 million acres of American agricultural land have been permanently conserved as of December 2022, comprising 42,000 parcels, the vast majority through conservation easement protections.

As America's major national agricultural land trust, the American Farmland Trust (AFT) has played an important role in this effort, conserving approximately 75,000 acres through its own protection efforts. However, AFT's primary role today is serving as an advocate and information center for farmland preservation practices and policy, while local and smaller regional land trusts hold the vast amount of farmland conservation easements.[13]

Farms Under Threat and the Farmland Information Center: Information as Preservation

A farmland preservationist could argue the most important tool of the last few decades isn't the rise of the conservation easement, or the proliferation of American land trusts, but the American Farmland Trust's "Farms Under Threat" report, published in 2022, along with the AFT's companion "The State of the States" report, which includes a state-by-state spatial analysis tool and policy scorecard.

"The 'Farms Under Threat' and 'State of the States' have been such useful tools because they definitely made many states and communities aware of the threat to their agricultural resources," says Jennifer Dempsey, the current head (as of early 2026) of American Farmland Trust's Farmland Information Center (FIC).

The origins of the FIC began with the FPPA, passed as part of the 1981 Farm Bill. The purpose of the act was to minimize the conversion of farmland to nonagricultural uses by federal programs, but a provision of the act also called for the establishment of a central information repository about state and local approaches to farmland preservation. By the early 1990s, however, the federal information repository still hadn't been established, so the Natural Resources Conservation Service (NRCS) provided financial resources to AFT to establish the FIC in 1995. Its purpose was to be an information clearinghouse on farmland preservation efforts and policies from across the nation. Today the online resource serves as a library for a wide variety of farmland preservation reports and research, some generated by FIC/AFT staff and some from other research and policy analysis sources.

A visit to the FIC's website (at www.farmlandinfo.org) reveals a literal encyclopedia of information about farmland preservation policy and practice, some of it buried deep within the website's information architecture, accessible only through targeted Internet searches. Dempsey says a key role of the FIC is to give national attention to local farmland preservation efforts.

"One of the things we (AFT) think is important about the Farmland Information Center is that it is the state and local governments that tend to innovate in farmland preservation programs in response to the threat of development in their local communities." As an example, she points to the creation of the first Purchase of Agricultural Conservation Easement (PACE) program in Suffolk County, New York, in the 1970s. PACE programs are described in more detail later in this chapter. "To share as many of those state and local innovations as possible is a real way to protect agricultural resources where they are," Dempsey says.

Visitors to the FIC website will find recently updated reports on a variety of farmland preservation initiatives, ranging from inventories of state and local PACE programs to a summary of the USDA's recently published "Census of Agriculture." Other recent resources take a more local and state-specific focus, such as a webinar on the economic impacts of farmland protection in Wisconsin.

But the "Farms Under Threat" report and its associated analytical tools are the FIC's most impactful research publications of the last decade. Cogently written and attractively designed, the report perhaps paints a bleak future for American farmland, but it also serves as a call to action for state and local farmland preservationists, who can point to the report's state-by-state and county-by-county analysis to galvanize local preservation efforts. The "Farms Under Threat" report began when FIC staff began thinking it was important to do a really comprehensive assessment of the nation's agricultural resources," explains Dempsey:

> We thought it would be very powerful to quantify the extent of agricultural land conversion and do it in a spatial manner. We felt other efforts to track land-use changes in America weren't focused on agricultural resources ... and we also wanted to understand if we were perhaps losing at a greater rate our most strategically important, most productive agricultural land.

Indeed, that concern proved true. Among other key findings, the report concluded that the nation's "Most Significant Agricultural Land"—its most versatile, productive, and resilient farmland and ranchland—will be over 50 percent more likely to be converted to development than other agricultural land during the report's forecast timeline.[14]

One of the most impactful parts of "Farms Under Threat" is its companion "State of the States" report, which evaluates each individual American state from several different analytical measures, ranging from specific types of development threats to conversion of the nationally significant farmland described above.[15] Another useful tool in the "Farms Under Threat" suite is its "Future Scenarios" projection, which utilizes Geographic Information System (GIS) technology to forecast farmland conversion to the year 2040, based on the tool's three growth models: "Business as Usual," "Runaway Sprawl," and "Better Built Cities." The tool can be employed on both a state and county level. For example, under the "Business as Usual" projection, Chatham County, where I work and live, is projected to lose an additional

14,700 acres by 2040. Under the Runaway Sprawl scenario, the projected loss is 19,400 during the same period. The tool also predicts that 44 percent of this conversion will take place on Chatham County farmland with AFT's "Most Significant Agricultural Land" designation, i.e. land with prime farming soils and other desirable agricultural characteristics.[16]

The "State of the State" report also includes a recently added state-by-state "Policy Scorecard" evaluating extent and effectiveness of each state's farmland preservation policy. "The policy scorecard evaluates each state across a suite of six state-level policies that we felt advanced farmland protection," explains Dempsey. The six policies were: 1. PACE programs; 2. Land-use planning and growth management policies; 3. Property tax relief for agricultural land; 4. Agricultural District programs; 5. Farm Link programs; and 6. State leasing programs.

> We wanted a place where state policymakers and farmland preservation advocates could go in and see the key elements of state-level policy approaches at a glance. Then, based on what they're looking for, they could identify a really good example of a farmland preservation statute or policy to replicate where they are.

According to Dempsey, one of the main policy benefits of online reports like "Farms Under Threat" and "The State of the States" are their comparative elements. "Together, these tools not only help identify the threats to farmland but also provide a broader understanding of the farmland preservation tools other places use and what they've been able to accomplish using them," she says. This can inspire state and local governments to adopt farmland preservation policies that have been successful elsewhere and help justify them to political boards and representatives.

As proud as Dempsey is of the "Farms Under Threat" report, she believes the farmland preservation movement is shifting to a greater focus on agricultural viability—keeping current farms financially solvent and providing opportunities to new farmers—and she plans to move the reporting focus of the Farmland Information Center with it.

> AFT's mission has always been about protecting farmland and encouraging sound agricultural management practices. But more recently in AFT's history, we have acknowledged the importance of agricultural viability, which is described in our mission as "Keeping farmers on the land." Obviously, a key way to protect agricultural land is make sure that farms and ranches are viable businesses.[17]

PACE Programs

If the Reagan administration's laissez-faire conservation policy pushed farmland preservation into the realm of nonprofit land trusts, many of America's individual states could no longer ignore the farmland conversion crisis impacting their rural communities and agriculture sectors by the late 1990s and early 2000s (or earlier in some northeastern states). This need has led to the proliferation of Purchase of Agricultural Conservation Easement, or PACE, programs, which provide government funds for the acquisition of agricultural conservation easements, typically through competitive grant programs to nonprofit land trust and other eligible organizations.

Conservation easement acquisition programs need not focus on agriculture. Some focus on lands associated with water quality, such as wetlands or riparian buffers, while others can focus on rare plant and wildlife habitats, natural and cultural resources, or other threatened ecosystems. When focused on agriculture, they compensate landowners for placing a conservation easement on their property, typically prioritizing the preservation of prime soil quality, the property's overall threat of development, and its future agricultural viability in the award selection process. Payment is generally based on the difference between the value of the land when restricted by the conservation easement and the "highest and best use" or "fair market value" (FMV). Many PACE programs require some sort of local matching contribution toward the conservation easement acquisition, which can be in the form of a complementary grant from a local source, private donations, the landowner "donating" some portion of the conservation easement in the form of uncompensated value, or often some combination of these three sources.

Some critics decry PACE programs as part of a neoliberal trend toward marketizing land preservation, with little expansive impact on farmland conversion at a landscape scale. These criticisms frequently evoke the conservation ethic of Aldo Leopold and the New Agrarians, arguing that market-based solutions such as conservation easement acquisitions can never inspire the widespread conservation practices and land stewardship values necessary for truly sustainable agricultural systems. With case law on conservation easement enforcement and amendments still in its infancy, others worry that state-funded conservation easements have an uncertain legal future. But advocates support conservation easements and PACE programs as a way for governments and their partners to work within America's "property rights" spirit of the law while still creating significant farmland preservation benefits. According to an AFT report (available on the Farmland Information Center), twenty-nine states had active state-level PACE programs as of January 2025, resulting in the protection of 3.81 million acres.[18]

And these figures do not include the contributions of local PACE programs. AFT estimated that at least 108 independently funded, local PACE programs operated in twenty-two states as of January 2025. These programs are typically operated by counties facing significant farmland conversion, although sometimes they are operated on a municipal or regional scale. AFT estimates local PACE programs have contributed to the protection of 1.17 million acres by acquiring or assisting in the acquisition of over 11,000 conservation easements at an overall expenditure of $2.3 billion in local funds. While funding for state PACE programs typically comes from an appropriation from the state's general budget, local PACE programs have much more diverse funding sources, ranging from local bond initiatives to real estate transfer fees to special sales tax designations.[19]

By definition, PACE programs focus on acquiring farmland conservation easements. But not all local and state grant programs focus on them exclusively, and some attempt to address the agricultural viability issue Dempsey described above. For example, North Carolina's PACE program, known as the Agricultural Development and Farmland Preservation (ADFP), also includes grants for local and regional agricultural infrastructure projects, such as training or distribution centers, as well as local farmland preservation planning efforts. In 2024, ADFP awarded approximately $1 million for agricultural infrastructure and development projects, including $487,326 for a new livestock arena and agriculture center and $450,000 for an animal science program in North Carolina communities. ADFP also awarded a total of $110,500 in 2024 to support farmland preservation planning processes in eight North Carolina counties. But the majority of its $18.2 million went toward conservation easement acquisition projects, awarded either to nonprofit land trusts or local governments.[20]

ACEP and Federal Farmland Preservation

Recognizing the rising interest in conservation easements, the United States Congress established a federal PACE program in 1996 originally called the Farmland Protection Program (FPP), operated under the auspices of the NRCS, the USDA's primary natural resource conservation division.[21] In 2002, NRCS changed the name of the FPP to the Farm and Ranch Lands Protection Program (FRPP), when Congress expanded the program's funding eligibility to include nonprofit land trusts. The program's name was changed again to ACEP-ALE in 2008 when that year's Farm Bill expanded the program's criteria to include grassland, pastureland, and certain types of forest land.

In 2020, American Farmland Trust conducted a comprehensive analysis of ACEP-ALE, including a survey of farmland owners who had participated in the program. Among a variety of findings, the study determined that 78 percent of parcels protected through ACEP-ALE and FRPP between 2001 and 2016 occurred in counties that fell above the state median for acres converted to urbanization. Many landowners who sold easements invested the proceeds back into their land either by buying more land, planting more acres, or adding livestock to their farm operation, including 65 percent of first-generation farmers who responded to the survey. Additionally, 48 percent of first-generation farmers paid off loans on land they already owned using ACEP-ALE/FRPP proceeds. "This evaluation shows that the FRPP protected threatened agricultural lands and kept it in active agricultural use, fulfilling its statutory purpose," AFT reported in its conclusions. "Results also show that the program spurred landowners to make changes that ensure a future for agriculture."[22]

Other farmland conservation programs offered by USDA/NRCS include the Wetland Reserve Program, which can be used to acquire both permanent and term (or temporary) conservation easements on privately held farmland or converted wetlands that can be successfully restored to protect water quality and wildlife habitat. Similarly, NRCS's Healthy Forest Reserves Program also funds acquisition of permanent and term easements to restore forestland that can promote the recovery of endangered or threatened species or provide other habitat and environmental benefits. NRCS's Regional Conservation Partnership Program (RCPP) is a partner-driven program approach that funds solutions to natural challenges on agricultural land, including funding for permanent conservation easements (typically through ACEP-ALE).

However, the vast majority of these federal conservation programs still promulgate the "rational conservation" ethos of incentivizing conservation practices through contract payments and cost share. Two of NRCS's most popular conservation programs are the Environmental Quality Incentive Program (EQIP or EQuIP) and the Conservation Stewardship Program (CSP). EQIP provides technical and financial assistance to farmers and forest owners to address natural resource management concerns such as improving water quality and reducing soil erosion and sedimentation. CSP offers annual payments based on implementation of a conservation plan developed in conjunction with NRCS representatives.[23] Although these programs are in the tradition of the Progressive and New Deal farmland preservation movement, they are focused on providing financial incentives for implementing conservation management practices and offer no permanent or temporary measures to prevent farmland conversion to higher density development.

New Horizons for Farmland Preservation

As farmland conservation easements proliferated in the 1990s and early 2000s, fueled by the rise of state, local, and federal PACE programs, the farmland preservation movement focused on the rational agrarian rhetoric of protected land resources. From this resource-based perspective, the conservation easement, with its focus on permanent land restrictions and built-in enforcement measures, is perhaps the ultimate farmland preservation tool. But conservation easements have always had their critics, and awareness is growing about their limitations. "Conservation easements are great tools for land that has limited human management and stewardship impacts," says Ian McSweeney. "But the more of those management needs you layer onto farmland, the more problematic conservation easements can be over time."

McSweeney is the co-executive director of the Farmers Land Trust, a nonprofit that embraces the farmland commons model for farmland protection and enhancing access to farmland. In the farmland commons model, farmland is owned by a nonprofit corporation, which establishes bylaws and procedures to govern the land's management and lease terms. "The nonprofit corporation holds the title to the farmland real estate and leases it out under secure, equitable, long-term lease structures," McSweeney explains.[24]

McSweeney is part of a growing cadre of farmland preservationists who recognize the enforcement and regulatory provisions within most conservation easements can prohibit the type of adaptive and innovative land uses that can keep farming viable in changing markets and social environments. Agrovoltaics, for example, are an adaptive farmland use. These are systems where agricultural land is used simultaneously for both solar power generation and crop or livestock production, but many conservation easements prohibit this type of adaptive, sustainable utility infrastructure.

Conservation easements can also ignore or challenge the economics that prevent access to farmland for many, especially American agriculture's traditionally disenfranchised classes—young, minority, and women farmers. Due to prevailing market conditions, even land permanently protected by conservation easements can be unaffordable for many farm operations, and their permanent restrictions may not be a good fit for entrepreneurial farmers seeking to live, grow food, husband livestock, and establish viable, adaptable farm enterprises. Conservation easements can also require financial resources and professional expertise for stewardship and conservation management practices that can be a barrier to landowners without access to financial capital or experienced land management advisors. And many farmers are skeptical about permanent restrictions that can limit future plans and options, even if they may not yet know what those plans and options may be.

This awareness has led to a new wave of farmland preservation initiatives focusing on creating access to farmland in competitive real estate markets, along with legal, economic, and social support for a new generation of American farmers. The farmland commons model championed by McSweeney is one example of this evolution. The Conservation Fund's "Farms Fund" program is another prominent example.

The Conservation Fund is a national conservation nonprofit that has long focused on entrepreneurial approaches to land preservation, often by buying large tracts of undeveloped land, then recouping their investment through partnerships with local and state grant programs. The organization's Working Farms Fund adapts this model to a smaller scale, by acquiring prime farmland under threat of development, recruiting tenants who want to establish or expand farming operations but lack the necessary capital to acquire the land themselves, then working with those tenants to craft a long-term plan to transfer ownership of the land as soon as the farm is financially stable. "What the farmers we work with need is not just loans but support in the transition to make their farms successful," explains Kristian Varsa, the director of the Farms Fund program. "A lot of these farmers are leasing on small acreage. They need a path to ownership, but there isn't a path to ownership that's affordable for them."[25]

These types of initiatives are often described as "Buy-Protect-Sell" (BPS) programs because they involve nonprofit or government agencies acquiring farmland, protecting it through conservation easements or other contractual agreements, and then eventually selling the land to farmers who would otherwise struggle to acquire farmland on their own. In the Farm Funds program, the "protect" component often comes through the sale of a conservation easement to a state, local, or federal PACE program. The farmland commons model relies on placing ownership of farmland and other farm infrastructure in the name of a nonprofit corporation whose bylaws and governing documents prioritize farmland preservation. Other initiatives are using crowdsource funding or Real Estate Investment Trust (REIT) models to achieve similar results. Steward is an investment company that uses an Internet crowdsource funding model to acquire farmland and other infrastructure for sustainable and progressive farming operations.

The Rural Advancement Foundation International (RAFI) adopts another approach to farmland preservation by providing financial counselling and advocacy to farmers who are struggling financially to hold onto their farmland, among other initiatives. The organization runs a financial crisis hotline and also offers an advocacy program to assist farm owners though a variety of financial and legal issues, from navigating federal financial assistance programs to mediation and natural disaster assistance. Meanwhile, nonprofit initiatives like the Young Farmers Network and AFT's "Women for the Land" program provide training, support, and advocacy for specific types and classes of farmers who have traditionally been excluded from farmland ownership or face significant structural barriers in America's conventional agriculture systems.

These are only a few examples of the American farmland preservation initiatives that are moving beyond the traditional tools of conservation easements and permanent land restrictions. Some American communities are implementing planning and zoning approaches; some are trying to help their farmers and farmland owners navigate the challenging process of farm transition planning. And hovering over it all is a persistent, if sometimes disparate, advocacy for the future of American farming. Call it Rational Agrarianism, Romantic Agrarianism, Resource Agrarianism, or New Agrarianism, the message is the same—American farmland and its farmers are worth preserving.

Notes

1 Yanhua Xie et al., "U.S. Farmland Under Threat of Urbanization: Future Development Scenarios to 20240," *Land* 12, no. 1 (2023), doi:10.3390/land12030574.
2 Xie et al., "U.S. Farmland."
3 "Production—United States," USDA Foreign Agricultural Service, https://www.fas.usda.gov/data/production/country/us.
4 Robert Whutnow, *The Left Behind: Decline and Rage in Small-Town* America (Princeton University Press, 2018), 6.
5 Nicholas F. Jacobs and Daniel M. Shea, *The Rural Voter: The Politics of Place and the Disuniting of America* (Columbia University Press, 2024), 373–374.

6 Don Stuart, *No Farms, No Food: Uniting Farmers and Environmentalists to Transform American Agriculture* (Island Press, 2022), 17–18, 40–41.
7 Elizabeth Byers and Karin Marchetti Ponte, *The Conservation Easement Handbook* (Land Trust Alliance, 2005), 7.
8 Federico Cheever and Nancy A. McLaughlin, "An Introduction to Conservation Easements in the United States: A Simple Concept and a Complicated Mosaic of Law," *Journal of Law, Property and Society* 1 (May 2014): 108–185.
9 Julie Ann Gustanski, "Protecting the Land: Conservation Easements, Voluntary Actions, and Private Lands." In Julie Ann Gustanski and Roderick H. Squires, Eds., *Protecting the Land: Conservation Easements Past, Present, and Future* (Island Press, 2000).
10 John L. Hollingshead, "Conservation Easements: A Flexible Tool for Land Preservation," *Environmental Lawyer* 3, no. 2 (February 1997): 319–362.
11 Richard Brewer, *Conservancy: The Land Trust Movement in America* (University Press of New England, 2003), 32.
12 "Gaining Ground," Land Trust Alliance, accessed August 29, 2024, https://landtrustalliance.org/land-trusts/gaining-ground/united-states.
13 "Land Protection Projects," American Farmland Trust, accessed August 29, 2024, https://farmland.org/project/land-protection-projects/.
14 "Farms Under Threat 2040: Choosing an Abundant Future," American Farmland Trust, https://farmlandinfo.org/publications/farms-under-threat-2040/.
15 J. Freedgood, M. Hunter, J. Dempsey, and A. Sorenson, "Farms Under Threat: The State of the States," American Farmland Trust, 2020, https://farmlandinfo.org/publications/farms-under-threat-the-state-of-the-states/.
16 "Farms Under Threat: Future Scenario Tools," American Farmland Trust, accessed June 7, 2025, https://development2040.farmland.org/.
17 Jennifer Dempsey (Director of the Farmland Information Center) in discussion with the author, March 27, 2025.
18 "Status of State PACE Programs," American Farmland Trust Farmland Information Center, https://farmlandinfo.org/wp-content/uploads/sites/2/2024/10/status-state-PACE-programs-2025.pdf.
19 American Farmland Trust Farmland Information Center, "Status of State PACE Programs."
20 "NCDA&CS Awards $18.2 Million in Farmland Preservation Grants," North Carolina Department of Agriculture & Consumer Services, September 13, 2024, https://www.ncagr.gov/news/press-releases/2024/09/13/ncdacs-awards-182-million-farmland-preservation-grants.
21 Craig Cox, "US Agriculture Conservation Policy and Programs: History, Trends, and Implications," *U.S. Agriculture Policy and the 2007 Farm Bill* (Woods Center for the Environment at Stanford University, 2006), 113–145.
22 Jennifer Dempsey, "Analyzing the Lasting Impacts of the Farm and Ranch Lands Protection Program," American Farmland Trust (August 2023), 12–13, https://farmlandinfo.org/wp-content/uploads/sites/2/2023/10/AFT-frpp-analyzing-lasting-impacts.pdf.
23 "Programs & Initiatives," Natural Resources Conservation Service/USDA, accessed February 17, 2025, https://www.nrcs.usda.gov/programs-initiatives.
24 Ian McSweeney (The Farmers Land Trust Co-Executive Director) in discussion with the author, May 23, 2025.
25 Krisztián Varsa (Director, Farms Fund) in discussion with the author, May 8, 2025.

Bibliography

American Farmland Trust Farmland Information Center. 2022. "Farms Under Threat 2040: Choosing an Abundant Future." https://farmlandinfo.org/publications/farms-under-threat-2040/.

American Farmland Trust Farmland Information Center. 2025. "Status of State PACE Programs." https://farmlandinfo.org/wp-content/uploads/sites/2/2024/10/status-state-PACE-programs-2025.pdf.

Brewer, Richard. 2003. *Conservancy: The Land Trust Movement in America*. University Press of New England.

Byers, Elizabeth, and Karin Marchetti Ponte. 2005. *The Conservation Easement Handbook*. Land Trust Alliance and Trust for Public Land.

Cheever, Frederico, and Nancy A. McLaughlin. 2014. "An Introduction to Conservation Easements in the United States: A Simple Concept and a Complicated Mosaic of Law," *Journal of Law, Property and Society* 1. doi:10.2139/ssrn.2650157.

Cox, Craig. 2006. "US Agriculture Conservation Policy and Programs: History, Trends, and Implications." In *U.S. Agriculture Policy and the 2007 Farm Bill*, edited by Kaush Arha, Tim Josling, Daniel A. Sumner, and Barton H. Thompson. Woods Center for the Environment at Stanford University.

Dempsey, Jennifer. 2023. "Analyzing the Lasting Impacts of the Farm and Ranch Lands Protection Program," American Farmland Trust. https://farmlandinfo.org/wp-content/uploads/sites/2/2023/10/AFT-frpp-analyzing-lasting-impacts.pdf.

Freedgood, J., M. Hunter, J. Dempsey, and A. Sorenson. 2020. "Farms Under Threat: The State of the States," American Farmland Trust. https://farmlandinfo.org/publications/farms-under-threat-the-state-of-the-states/.

Gustanski, Julie Ann. 2000. "Protecting the Land: Conservation Easements, Voluntary Actions, and Private Lands." In *Protecting the Land: Conservation Easements Past, Present, and Future*, edited by Julie Ann Gustanski and Roderick H. Squires. Island Press.

Hollingshead, John L. 1997. "Conservation Easements: A Flexible Tool for Land Preservation," *Environmental Lawyer* 3, no. 2.

Jacobs, Nicholas F., and Daniel M. Shea. 2024. *The Rural Voter: The Politics of Place and the Disuniting of America*. Columbia University Press.

Land Trust Alliance. n.d. "Gaining Ground." https://landtrustalliance.org/land-trusts/gaining-ground/united-states.

Stuart, Don. 2022. *No Farms, No Food: Uniting Farmers and Environmentalists to Transform American Agriculture*. Island Press.

Whutnow, Robert. 2018. *The Left Behind: Decline and Rage in Small-Town America*. Princeton University Press.

7 Conservation Easements

"Land conservation, like jazz, is an American invention," writes Richard Brewer in his book *Conservancy: The Land Trust Movement in America.* [1] And if conservation is like jazz, the conservation easement is its *Kind of Blue*, the 1959 Miles Davis album widely regarded as the most important jazz recording of all time.

And like *Kind of Blue*—a smoky, bluesy concoction unlike any other jazz recording before or since—the conservation easement remains somewhat mysterious, its mechanics often as little understood by the public at large as its appeal. Already you know the American spirit of the law favors land-use policies that work within its bias toward individual property rights. The previous chapter described how conservation easements have been used to permanently preserve and protect approximately 7 million acres of farmland,[2] a small but not insignificant portion of the over 61 million American acres of all types of land preserved through conservation easements as of 2020.[3] And you've read how more and more state and local governments are putting precious public dollars toward the acquisition of conservation easements through Purchase of Agricultural Conservation Easement (PACE) and similar grant programs, partnering closely with America's robust land trust movement.

But if conservation easements have become the American land conservation movement's most prominent tool, challenges remain for their widespread application. Put simply, a conservation easement is a voluntary legal agreement or deed restriction that limits future activities on land to protect its conservation values.[4] Most conservation easements are long and complex legal documents, often misunderstood even by the land trust professionals who attempt to explain them to landowners. In trying to protect against all possible future land development contingencies, they frequently limit or exclude emerging sustainable practices, such as wind and solar power generation. They typically require expensive transaction costs such as surveys and appraisals that are prohibitive for many agricultural landowners of modest financial means. And perhaps most importantly, their long-term efficacy is in many ways yet to be evaluated. As more and more of the 61 million acres already protected through conservation easements passes from one generation to the next, or

DOI: 10.4324/9781003644545-9

out of family hands completely, the next challenge to the American land trust community will not be acquiring new conservation easements but defending the ones they already have.

Despite these challenges, conservation easements are here to stay. And to understand farmland preservation in American, one most understand conservation easements. And so, let's meet the Browns, a hypothetical farming family in a hypothetical farming community considering a conservation easement to preserve their farmland.

How Conservation Easements Work: Meet the Browns

Mr. and Mrs. John Brown own a 100-acre tract of land in a rapidly urbanizing county in central North Carolina. The land was part of a larger 1,000-acre tract that was actively farmed by Brown's grandfather but was subdivided among Grandfather Brown's heirs when he died. Mr. Brown's father worked in a factory but always kept forty to fifty head of cattle on his share of the family farmland as a side business and built a two-story house on the property, the home where Mr. Brown grew up. After Mr. Brown's parents died, he inherited the entire 100-acre tract, there being no other heirs, and remodeled the house for his wife and only child, Cynthia. Like his father, Mr. Brown never worked full time as a farmer. He spent his career as an accountant for a local trucking company and raised cattle as a side business. Although Mr. Brown doesn't consider himself an environmentalist, he takes pride in managing the farm sustainably, keeping his livestock out of perennial streams, where their waste can pollute water supplies; allowing the wooded portions of his property to grow into a mature hardwood forest, providing excellent habitat for birds, reptiles, and small mammals; and managing his pasture areas on a rotational grazing schedule without use of pesticides or herbicides whenever possible.

The farm comprises a single, rectangular tax parcel. The Browns' house and an assortment of farming-related structures (barns, sheds, a silo, etc.) encompass a 5-acre portion of the property we will call the "building envelope," or in some cases is called the "farmstead envelope." This envelope is on the southern boundary of the property, which also abuts the county road the Browns use to access their property. On the "back" of the farm is approximately 20 acres that is forested and abuts a perennial stream that comprises the farm's northern boundary line. The remaining 75 acres are fields used for pasture and hay. Although Mrs. Brown operates a significant vegetable garden for the couple's personal use within the 5-acre envelope, the sole income-producing activity of the farm is raising cattle for small-scale beef production.

Cynthia is married, with two children, and lives in the nearby city. Cynthia has never been interested in being a farmer or living on the family farm, although the farm's real estate is the Browns' only significant asset, which they plan to leave to Cynthia when they die. This is not an insignificant bequest, for this area is urbanizing rapidly, and undeveloped land here is currently valued at $20,000 per acre. The Browns' modest farmhouse is worth another $200,000, although, realistically, anyone who purchased the property would be buying it for residential development and most likely would bulldoze the house and the farm structures as part of their development plans.

Both Mr. and Mrs. Brown are in their seventies and remember a time when the surrounding area was a thriving farm community. Mrs. Brown grew up just a few miles away, also on a small-scale cattle farm, and the Browns were high school sweethearts. As they have gotten older, they have grown nostalgic about the sense of community and fellowship they enjoyed in their younger years, and proud also of the farming legacy they have carried on from their parents and grandparents. The Browns recently met with a representative from the local land trust, who told them about a state grant program that would pay them for a conservation easement on their property. After consulting with their daughter, who is amenable to the conservation easement (although she doesn't fully understand what it means), the Browns agree to work with the land trust on the conservation easement, provided they can get the state grant.

The Legal Framework

One of the reasons conservation easements have gained popularity is because they work within both America's property rights "spirit of the law" and the related "bundle of rights" framework introduced in Chapter 3.

Indeed, the Browns have already conveyed some of their farmland rights within this "bundle." A neighbor holds a lease granting the right to hunt their property during deer and turkey seasons. Also, near the rear of the property, bisecting its northeast corner, is a right-of-way easement Mr. Brown's parents conveyed to the local power company for a transmission line. Although the Browns still own this part of their property, the power company has extensive rights to manage the right-of-way area for its power infrastructure, with little the Browns can do or complain about it.

Although both of these are examples of the "bundle of rights," they function differently. In legal terms, the hunting lease is a "leasehold estate," in which a property owner or "lessor" grants the tenant or "lessee" the right to use property in a specific way for a specific amount of time in exchange for some consideration, usually a rent or sum of money. In contrast, the power company right-of-way is called an easement in gross because it grants access to the power company (or any other specific entity) for a specific purpose (to maintain the power lines) and is not directly tied to a specific piece of land that benefits from the access. In contrast, an easement appurtenant is a legal right that burdens one piece of land for the benefit of another, usually an adjacent piece of land, such as an easement that allows access from a public road to one property over another property.[5]

The Browns also have a mortgage on their property of $50,000. This was a twenty-year loan of $100,000 the Browns took out ten years ago to make some improvements to their home and farm buildings and have paid down over time. This mortgage is yet another type of property right known as a security interest, or a legal right acquired in one's property by a creditor, or lienholder. A lien usually prohibits the sale of a property until the creditor's obligation is satisfied, as is the case with the Browns' mortgage.[6]

In legal terms, a conservation easement is an easement in gross, because it extracts certain rights from the landowner's bundle of rights for a specific entity, i.e. the land trust, but not for the benefit of an adjacent piece of property. In the past, American common law has not looked favorably on easements in gross. Prior to modern deed recording systems, courts were concerned that easements in gross would be forgotten or overlooked, then rediscovered decades later to the torment of the current property owner. Thankfully, computerized deed searches have helped ease this concern.[7]

Conservation easements are also deeds of conveyance, a legal instrument through which property owners transfer away their right to undertake specified activities on their land. For conservation easements, for example, these transferred rights could be the right to subdivide, build houses, or practice industrial manufacturing on the property. This means that, in order for a conservation easement conveyance to occur, some entity must agree to hold those "rights" in trust and, if necessary, defend their restrictions in a court of law. In effect, the conservation easement holder has been granted the right to enforce the grantor's promise not to engage in those uses.[8]

In addition, conservation easements function as "negative" easements, restricting what the landowner may do on that parcel. Like easements in gross, the American legal system has historically viewed "negative" easements skeptically, preferring not to encumber the rights and economic opportunities of future landowners based on the personal choices of a previous one. Passage of the Uniform Conservation Easement Act (UCEA) in 1981 helped to address these legal concerns, but there remains a prejudice against conservation easements in the American spirit of the law that could rear its head in generations to come.[9]

Like other easements in gross, conservation easements may be conferred for a term or "in perpetuity." Term easements expire at the end of a specified period, while perpetual easements, in theory, never expire, except by judicial order. Although some government programs require or support term conservation easements, most nonprofit land trusts require perpetual easements. In part, this is to benefit their nonprofit status. If the mission of a land trust is to permanently protect land for future generations, only a perpetual easement reasonably strives toward this mission. Furthermore, the due diligence and transaction expense for a term easement and perpetual easement is essentially the same, so many land trusts prefer not to invest their precious fiscal assets and capacity into a conservation easement that will go away after a set number of years.

Although landowners are generally more amenable to term conservation easements than land trusts, the federal tax incentives offered for conservation easement donations only apply to perpetual easements. And many of the grant programs that fund conservation easement acquisitions will only pay for perpetual conservation easements. This isn't a problem for the Browns, however, because they want to see their land and its connection to their family legacy protected permanently. For them, a conservation easement only makes sense if it protects their land for their daughter's generation and generations beyond.

But preserving family legacy isn't a valid legal reason for granting a conservation easement. Under the UCEA, a conservation easement may be created to meet the following requirements: (1) retain or protect natural, scenic, or open-space values of real property or assure its availability for agricultural, forest, recreational, or open-space use; (2) protect natural resources; (3) maintain or enhance air or water quality; or (4) preserve the historical, architectural, archaeological, or cultural aspects of real property.[10] Today, all states have adopted the UCEA (Wyoming was the last holdout but passed adoption in 2005), and many states have adopted their own tax incentives and other benefits to encourage conservation easements. This has led to a mosaic of state and federal laws created to support or influence the conservation easement conveyances throughout the United States, most referring to the conservation easement's benefits in terms of its preservation of natural and historic benefits, the protection of prime soils, for example, or habitat for rare and endangered species. Social benefits, such as the farmland's associated symbolic and the Browns' farming legacy in the community, are notably *not* valid legal purposes for a conservation easement, even if the easement preserves these things as well.[11] The Browns' land trust representative explains that reference to this mosaic of laws and supporting legal statutes specific to North Carolina will be explicitly stated in the "Purpose" section of the conservation easement. But she assures them that, aside from the legalese, the conservation easement can achieve their goals for the property.

What Conservation Easements Mean for Landowners

The reasons that landowners like the Browns decide to convey perpetual conservation easements vary widely. If the Browns' primary motivation is to protect their family legacy of farming in a rapidly urbanizing area, a secondary reason for the Browns is the sense of stewardship and environmental sustainability they have always rendered toward their property. They believe the conservation easement will help impose that same stewardship ethic on future owners. Although the Browns are not financially motivated, they are of modest financial means, so a third motivation is the economic incentives available to them for a conservation easement conveyance. A conservation easement works well for them because their children have little economic incentive for keeping their land unrestricted, share in the desire to preserve the family legacy of farmland ownership, and are generally supportive of their parent's estate planning goals.

Other landowners may be more financially motivated, viewing the sale of a conservation easement as an opportunity to generate income from their property without major changes in the way they manage or use it. Still others may use conservation easements as an estate planning tool, reducing the value of their property in an attempt to avoid estate taxes, create an equitable distribution of assets among heirs, or both. A study published in 2020 attempted to categorize landowner motivations for participation in perpetual conservation easements by synthesizing forty-three previous studies. Sponsored by the habitat conservation group Ducks Unlimited, this study determined that successful information transfer by a technical advisor (i.e. the land trust representative who meets with the Browns) or a member of the landowner's social network was one of the key drivers of a conservation easement conveyance. "Successful acceptance of a program or conservation message is more likely if communication is conducted by someone in the same social network as the target audience," the study reported. Economic incentive was also a primary reason, although not necessarily an expectation for financial gain, but also the financial obligation or cost the landowner will incur in a conservation easement conveyance, and how the land trust or government programs could help offset or cover those expenses.

Interestingly, the study did not identify ecological responsibility (harkening back to Leopold's "land ethic") as a primary landowner motivation for conveying a conservation easement but did find a significant correlation with sense of place. "Sense of place, for which we observed positive trends, is driven by the meanings and connections individuals develop with their environment," the study states. As noted in Chapter 1, sense of place is a term that describes the meanings and attachment to a setting held by an individual or group. We shall return to the social and place dimensions of farmland preservation later in the book. Ultimately, the study recommended that those seeking to promote conservation easement focus on increased landowner interactions at multiple scales of governance structures, emphasizing the social dimensions of landowner motivations over economic and ecological ones.[12]

Writing over twenty years ago, Richard Brewer developed an alternative hypothesis in his book on the land trust movement. Summarizing his own personal experience with the research of the day, Brewer states: "All the lines of evidence seem to converge on the land trust folk principle: As a general rule, people donate property for preservation out of a love for the land; financial considerations are not unimportant but are secondary."[13]

Stewardship and Legal Defense

Conservation easements put significant restrictions on the way land can be managed, developed, and subdivided in perpetuity. But a conservation easement is only as strong as the mechanisms it puts in place to monitor the land use of future landowners and, if necessary, take legal action to defend the conservation easement in court.

To achieve the legal tools necessary for this perpetual protection, conservation easements convey to the easement holder, i.e. the land trust, the right to monitor the property for potential violations. For most land trusts, this means conducting a site visit at least once a year (although with the advent of satellite photography and drone technology, some larger conservation easements are now monitored from the air, not a physical visit).

Although annual monitoring is by no means a comprehensive way to monitor a property—even with aerial monitoring, some violations can go unnoticed, especially on larger properties—it remains the standard practice among land conservation organizations, often supplemented by published property transfer notices and local gossip. A change in ownership is a vulnerable time for a conservation easement. A competent closing attorney should find the conservation easement in a title search, and inform their clients of its terms and restrictions, but this information sometimes slips through the cracks of a real estate transaction, or is ignored deliberately, making the exchange of title a strategic window for land trusts to contact the new landowners, establish relations and communications, and inform them about the conservation easement restrictions and how it will impact their day-to-day use of the property.

Typical easements violations include new driveways or farm roads, a prohibited building or structure, or a prohibited timber harvest. In many cases, the easement holder/land trust can negotiate a remediation for the violation with the landowner. However, some easement violations are so significant they cannot be mediated, and in these cases the easement holder is obliged to litigate the easement violation in court. To prepare for these situations, most land trusts have a legal defense fund (and land trusts accredited through the Land Trust Alliance's accreditation program must meet strict legal defense fund requirements). Land trusts typically request that the landowner contributes to this fund as part of the conservation easement conveyance. Land trusts who are members of the Land Trust Alliance and meet its requirements are also eligible to participate in an insurance program for conservation easement legal defense that will help offset legal fees in an adjudicated easement violation case.

However, these provisions assume the conservation easement holder has not only the legal means to enforce it but also the organizational capacity. Given the perpetual nature of most conservation easements, this is no insignificant responsibility, forcing both landowners and land trusts to seriously consider the easement holder's long-term viability. To address these considerations, many conservation easements include provisions for a "back-up" easement holder or co-holder, who can assume these enforcement responsibilities should something occur to the original conservation easement holder. However, given the rather brief duration of conservation easement history in America, long-term conservation easement stewardship and legal defense practices remain largely untested. Only time will tell if legal defense funds, conservation easement insurance, and back-up easement holders are sufficient to meet the future threats to conservation easement protections.

PACEs

When the Browns meet with their land trust representative, they are informed about North Carolina's Agricultural Development and Farmland Preservation Trust Fund, a competitive grant program for agricultural conservation easement acquisitions run by North Carolina's Department of Agriculture and Consumer Services. This program is commonly called by the acronym ADFP and is an example of a Purchase of Agricultural Conservation Easement, or PACE, program. The terms and conditions for PACE programs differ widely, so the ADFP is discussed here only as an example.

To apply for ADFP funding, landowners must work through a qualified organization, typically a nonprofit land trust, although government agencies such as a local Soil & Water Conservation District are also eligible. This organization will apply for the ADFP grant on the Browns' behalf and, if received, manage the terms and conditions of the conservation easement acquisition so that the landowner can receive financial compensation when the conservation easement is conveyed. The ADFP will supply payment for up to 50 percent of the conservation easement value to the landowner, with the landowner expected to "donate" the remaining 50 percent or seek compensation for the remaining value from other local or federal grant sources.

To be eligible for ADFP funding, the Browns must agree to restrict the use of the majority of their property to agriculture in perpetuity through an agricultural conservation easement. They can continue to use the property for customary rural enterprises, including agritourism and selling farm products, but sacrifice the right to use it for larger residential, commercial, or industrial purposes. Some conservation easements permit limited additional homesites, but in the Browns' case, the conservation easement limits residential development to the current homesite, which they can modify or improve up to a limited square footage or replace with a newer home of roughly the same size if they want, or rebuild in the case of a fire or natural disaster.

New farming infrastructure, such as additional barns and farm-related outbuildings, will be limited to the existing Farmstead Building Envelope, whose area and total acreage will be defined in the conservation easement document and delineated by a survey. Furthermore, the conservation easement will restrict the total amount of impervious surface (roads, driveways, roofs, etc.), and there can be no mining or subsurface alteration on the property after the conservation easement is in place. Although small-scale solar and windmills are permitted for personal use, provided they fall within the impervious surface limit, large-scale alternative energy production such as solar or wind farms are not allowed.

Since PACE programs pay landowners for conservation easements, they typically get to dictate their terms. The land trust representative tells the Browns that, if they wish to forgo the state funding opportunity, the restrictions outlined in the conservation easement can be less prohibitive. In this case, the Browns could still donate the conservation easement to the land trust. This means the Browns would not be financially compensated for their conservation easement, but they would still have the permanent development restrictions placed on their property and the peace of mind of protecting their family legacy of land ownership. And they are still eligible for tax benefits associated with qualified conservation easement donations. For donated easements, the land trust's internal guidelines permit one additional house per 50 acres, so the Browns could retain the right to build two additional homes on their property—perhaps one for their daughter and one marketable lot they could sell to help finance their retirement. Other provisions regarding impervious surface and additional farming structures are also typically less restrictive when working with donated easements. However, if they want to pursue the state funding, they will have to abide by the state program's easement terms and requirements. The land trust representative also tells the Browns that the funding for ADFP grants is extremely competitive, and there is no guarantee they will get PACE funding if they apply. Finally, she advises them that, if successful, PACE funding will likely extend the timeline for their conservation easement transaction for several years due to increased contract and transaction due diligence required in the PACE grant contract.

Benefits, Drawbacks, and Future Considerations

By intentional design, the conservation easement appears to be a good option for our hypothetical Brown family. Primarily, it is a tool that allows the Browns to preserve their family legacy of farming, with the assurance that the land trust will serve as a long-term, perhaps indefinite, steward and defender of their land and conservation values. In return, the Browns can continue to use, manage, and live on their land much as they always have, with only the relatively minor inconvenience of an annual visit from their land trust representative. The Browns also like the idea that their land can continue to function as farmland, serving as a monument to their community's agrarian past, even as the surrounding area continues to develop around them.

Though they are not wealthy, the Browns can enjoy some of the financial incentives in place for conservation easement conveyances. Depending on the financial terms of the conservation easement conveyance, they may qualify for a federal income tax deduction, significantly minimizing their federal income tax over the next several years. If they are able to receive a PACE grant, they will even receive money they can use to do some much-needed maintenance and improvement on their home and put some aside for their children and grandchildren's inheritance.

But the conservation easement is not free. A conservation easement conveyance is like any other transfer of real estate rights, and may require a survey, an appraisal, and a title policy. Most conservation easements today require an environmental site assessment to inspect and identify any contamination or toxic pollution problems on the property. The land trust may require the Browns to pay them a service fee to cover their incurred time and expenses, as well as a contribution to the land trust's legal defense fund.

Depending on the size and complexity of the survey, these out-of-pocket transaction costs can easily reach into the tens of thousands, which is beyond the financial capability of many landowners. Aware of this burden, many land trusts try to offset these landowner expenses through grants or deferred gifts made through a will or estate plan. Still, it is not uncommon for a landowner to incur some upfront, out-of-pocket expense for a conservation easement conveyance, even if they will eventually be reimbursed for some or all of that expense through a PACE grant or other sources.

One issue the Browns must address if they want to place a conservation easement on their property is their current mortgage of $50,000. Since the mortgage functions as a lien on the title to the property, it could threaten the enforceability of the conservation easement. If the lender ever foreclosed on the property, the conservation easement could be extinguished because the bank's lien precedes the conservation easement restrictions in the property's chain of title. To address this issue, the Browns could (a) pay off the mortgage with the proceeds of the PACE funds, eliminating the mortgage's lien on the property's title; or (b) ask their lender to subordinate the mortgage to the conservation easement. "Subordination" is a term that means the mortgage lender will sign a legal document allowing the conservation easement to be first in the chain of title, so that in the event of foreclosure, the integrity of the easement remains intact. A mortgage subordination is also a requirement in order for the donor of a conservation easement to qualify for federal income tax benefits associated with their conservation easement conveyance. The Browns will have to address this issue with their mortgage lender, the land trust representative tells them. Some lenders are familiar with subordination requests for conservation easements and are willing to subordinate their mortgage provided the balance of the loan is far below the market value of the property, as in the Browns' case. However, each lender approaches subordination requests differently, so the Browns should initiate this discussion with their bank as soon as possible.[14]

Any document that by design attempts to anticipate and restrict future uses of a property will inevitably fail in some measure, and one of the biggest drawbacks of conservation easements from a landowner perspective is its lack of adaptability. Of course, this rigidness is the very aspect of conservation easements that allows it to preserve and protect land from future development, but in practice it might unintentionally limit future uses of the property that can maintain its agricultural viability.

Take, for example, the issue of solar and wind energy. Many conservation easements of the past and present either prohibit or severely restrict alternative energy production due to the impacts of the infrastructure required to support it. True, installing a solar farm on one of the Browns' pastures would require substantial infrastructure and significantly increase impervious surface, creating environmental impacts in the local watershed. However, alternative energy technology is advancing rapidly, and already there are substantial advances in agrovoltaics, which describes solar energy infrastructure that permits both agriculture and solar energy production. In a world where small- and medium-scale farmers face increasingly thin profit margins, agrovoltaics might prove a sustainable economic model for their future business, yet a conservation easement may prohibit this opportunity.

An even more serious issue for most landowners is the way that a conservation easement limits how land can be inherited and conveyed to heirs. Already we have said that the Browns' conservation easement will limit future subdivision. That isn't an issue for the Browns, since they only have one heir, their daughter. However, let us now give the Browns two heirs, a son and a daughter. Without subdivision, the Browns' two children will have to inherit the land as a joint tenancy, which refers to a legal arrangement where two or more people own a property together, each with equal rights and obligations. In the present, the Browns' children may be amenable to this condition, because they want to honor their parents' wishes to see the land preserved. However, needs and situations change, and in the future, their joint tenancy arrangement may lead to disagreements and conflicts, particularly if the financial needs of one or both of their children change substantially in the future.

Public Benefits and Drawbacks

For the public, the benefits of conservation easements are perhaps less obvious. One reason for their widespread adoption is that they provide a voluntary, nonregulatory option for permanent and significant natural resource conservation and protection. Conservation easements work within the "bundle of rights" framework, and they allow ecologically minded landowners the opportunity to make a personal commitment to permanently protecting their property. In a country where land use regulation threatens the spirit of the law, American governments like this voluntary aspect of conservation easements because they do not require restrictive land-use regulations. But critics argue that the personal-choice dimension of conservation easements leads to a "patchwork protection" approach that creates the illusion of government advocacy without the broadscale benefits that can truly lead to landscape-scale environmental and farmland preservation benefits.

A more pressing concern, for our current age, are equity considerations. Critics such as the environmental scholar Dorceta Taylor argue that, for much of its history, the American conservation movement has been dominated by an affluent, White class with a long history of racial bias and discrimination.[15] Although significant effort in recent decades has been made to overcome this history, structural barriers remain for ethnic and minority landowners who want to preserve their land through conservation easements. One is the issue of "heirs property," a situation where family land is owned jointly as tenants in common by the descendants of a deceased person whose estate did not clear probate. In essence, this means that the property has been inherited as a joint tenancy, perhaps for several generations, creating many different owners with essentially equal rights to manage and sell the property. For heirs property owners, it can be almost impossible to reach consensus about conveying a conservation easement or making other long-term decisions regarding the future of their land. And in some cases, unscrupulous investors can use real estate laws to force the sale of an heirs property by acquiring a share of this joint ownership from an unsuspecting heir. Unfortunately, this situation frequently, though by no means exclusively, affects African American landowners in America, limiting their ability to protect and preserve their land through conservation easements. And even landowners of modest means like the Browns may struggle to pay the upfront and out-of-pocket expenses conservation easements typically require or procure competent legal and financial counsel during the process. Sadly, these challenges, and others like them, can put conservation easements out of reach for all but affluent landowners.

Finally, critics have argued that conservation easements encourage a spirit of the law that discourages government involvement in meaningful land use protection. Property rights scholar Eric Freyfogle is a strong advocate of this perspective. Freyfogle argues that landowners should not be paid, or receive financial compensation, for conservation easements. "Payment programs lend support to the claim that the activities being avoided are, and ought to be, legal," he argues.[16] Legal scholars Nancy McLaughlin and John Cheever echo Freyfogle's objections to conservation easements, arguing that providing financial incentives to private landowners to encourage them to protect their land saps the will and ability of governments to regulate land use.[17]

Such arguments remain in the minority, however, as conservation easements continue to proliferate in American land use policy, along with government programs that support their acquisition. The only thing clear at this point is that the history of conservation easements as a primary tool for farmland preservation is not yet long enough to provide clear answers. For now, the "bundle of rights" approach to farmland preservation has taken hold, with hopes that a communal spirit toward land stewardship eventually follows. It is a useful and proactive approach to farmland preservation, adopting the American spirit of the law—if not quite within it—but conservation easements are far from the only way to preserve a farm.

Notes

1 Richard Brewer, *Conservancy: The Land Trust Movement in America* (University Press of New England, 2003), 1.
2 "Protected Agricultural Lands Database," AFT Farmland Information Center, accessed August 19, 2024, https://farmlandinfo.org/statistics/pald/.
3 "61 Million Acres Voluntarily Conserved in America," Land Trust Alliance, December 2, 2021, https://landtrustalliance.org/newsroom/press-releases/61-million-acres-voluntarily-conserved-in-america-2020-national-land-trust-census-report-reveals.
4 National Conservation Easement Database, "What Is a Conservation Easement," accessed August 28, 2024, https://www.conservationeasement.us/what-is-a-conservation-easement/.
5 Mark P. Thompson, *Modern Land Law* (5th edition) (Oxford University Press, 2012).
6 "Lien," Cornell Laws School Legal Information Institute, July 2023, https://www.law.cornell.edu/wex/lien#:~:text=Legal%20Information%20Institute-,lien,the%20type%20of%20creditor%20involved.
7 Brewer, *Conservancy*, 148–149.
8 J. Owley, F. Cheever, A.R. Rissman, M.R. Shaw, B.H. Thompson Jr., and W.W. Weeks, "Climate Change Challenges for Land Conservation: Rethinking Conservation Easements, Strategies, and Tools," *Denver Law Review* 95 (2017): 727.
9 John L. Hollingshead, "Conservation Easements: A Flexible Tool for Land Preservation," *Environmental Law* 3 (1996): 319–361.
10 Uniform Conservation Easement Act, 1981, accessed September 2, 2024, https://landconservationnetwork.org/wp-content/uploads/legacy-files/Uniform%20Conservation%20Easement%20Act-1.pdf.
11 Federico Cheever and Nancy A. McLaughlin, "An Introduction to Conservation Easements in the United States: A Simple Concept and a Complicated Mosaic of Law," *Journal of Law, Property and Society* 1 (May 2014): 114.
12 Kaylan Kemink, Vannesa Adams, Robert Pressey, and Johann Walker, "A Synthesis of Knowledge about the Motives for Participation in Perpetual Conservation Easements," *Conservation Science and Practice* 3, no. 2 (2021): e323.
13 Brewer, *Conservancy*, 160.
14 "Mortgage Subordination Fact Sheet," Land Trust Alliance, accessed June 9, 2025, https://projects.sare.org/media/pdf/9/6/2/962193appendix-4-mortgage-subordination.pdf.
15 Dorceta E. Taylor, "The Rise of the Environmental Justice Paradigm: Injustice Framing and the Social Construction of Environmental Discourses," *American Behavioral Scientist* 43, no. 4 (Jan. 2000): 528–531.
16 Eric Freyfogle, *On Private Property: Finding Common Ground on the Ownership of Land* (Beacon Press, 2007), 129.
17 Frederico Cheever and Nancy A. McLaughlin, "Why Environmental Lawyers Should Know (And Care) About Land Trusts," *Environmental Law Reporter News & Analysis* 10223 (2004).

Bibliography

American Farmland Trust Farmland Information Center. n.d. "Protected Agricultural Lands Database." https://farmlandinfo.org/statistics/pald/.

Brewer, Richard. 2003. *Conservancy: The Land Trust Movement in America*. University Press of New England.

Cheever, Frederico, and Nancy A. McLaughlin. 2004. "Why Environmental Lawyers Should Know (and Care) about Land Trusts," *Environmental Law Reporter News & Analysis* 10223.

Cheever, Frederico, and Nancy A. McLaughlin. 2014. "An Introduction to Conservation Easements in the United States: A Simple Concept and a Complicated Mosaic of Law," *Journal of Law, Property and Society* 1. doi:10.2139/ssrn.2650157.

Freyfogle, Eric. 2007. *On Private Property: Finding Common Ground on the Ownership of Land*. Beacon Press.

Hollingshead, John L. 1996. "Conservation Easements: A Flexible Tool for Land Preservation," *Environmental Law* 3.

Kemink, Kaylan, Vannesa Adams, Robert Pressey, and Johann Walker. 2021. "A Synthesis of Knowledge About the Motives for Participation in Perpetual Conservation Easements," *Conservation Science and Practice* 3, no. 2. doi:10.1111/csp2.323.

National Conservation Easement Database. n.d. "What Is a Conservation Easement." https://www.conservationeasement.us/what-is-a-conservation-easement/.

Owley, J., F. Cheever, A.R. Rissman, M.R. Shaw, B.H.ThompsonJr., and W.W. Weeks. 2017. "Climate Change Challenges for Land Conservation: Rethinking Conservation Easements, Strategies, and Tools," *Denver Law Review* 95: 727. doi:10.2139/ssrn.3218525.

Taylor, Dorceta E. 2000. "The Rise of the Environmental Justice Paradigm: Injustice Framing and the Social Construction of Environmental Discourses," *American Behavioral Scientist* 43, no. 4. doi:10.1177/0002764200043004003.

Thompson, Mark P. 2012. *Modern Land Law*, 5th edition. Oxford University Press.

Uniform Conservation Easement Act. 1981. *Uniform Conservation Easement Act*. https://landconservationnetwork.org/wp-content/uploads/legacy-files/Uniform%20Conservation%20Easement%20Act-1.pdf.

8 Zoning, Planning, TDR, and Agricultural District Programs

If conservation easements work within the American spirit of the law, create permanent and legally enforceable farmland protections, and spur the nation's robust land trust movement, they will never possess the capacity necessary to broadly address America's farmland conversion issue. Conservation easements can be expensive and heavy-handed. They lack adaptability and foresight. They complicate estate distribution and inheritance. Some might argue their legality hasn't yet been definitively established by the American court system. Above all, they are a farm-by-farm answer to a problem that requires solutions at a landscape scale.

These limitations necessarily constrain conservation easements as a primary tool for American farmland preservation. However, America's legal and cultural emphasis on individual property rights largely prohibits extensive regulatory approaches. These dichotomies push American farmland preservation policy into the realm of planning and zoning, typically at the level of county and municipal governments, where the tension between individual and social rights frequently leads to uniquely American strategies. These creative solutions are part of a "quieter revolution" of place-based, public–private, multistakeholder approaches to farmland preservation and other American land-use challenges.

The following two chapters introduce a variety of these zoning, planning, and market-based approaches. This chapter introduces the concepts of Agricultural Protection Zoning (APZ) and Agricultural District Programs (ADPs), which use special property designations to preserve and promote farmland preservation in areas that still have robust farm communities. This chapter also reviews programs that attempt to monetize land conservation into market-based credit systems, such as Transfer of Development Right (TDR) programs and "in-lieu fees." The discussion continues in the following chapter when we look at other state and local initiatives such as Greenbelt Programs and public bonds. But first is a brief introduction to the American tradition of zoning and planning.

The Rise of Zoning and Planning

As America became more urbanized during the Industrial Revolution, and Manifest Destiny pushed settlers to the extent of her frontiers, the Founders'

DOI: 10.4324/9781003644545-10

vision of "inalienable" property rights faced an inevitable reconceptualization. The young nation was still primarily rural and favored the laissez-faire property rights perspective of Blackstone and Locke, but urban areas could no longer afford to let the market completely dictate their land uses.[1]

Between 1800 and 1900, America's urban inhabitants increased from 6 to 40 percent of the country's overall population, and cities like Chicago exploded, growing from a population of 30,000 in 1850 to over a million by 1890. Competition for space within these urban landscapes became intense, with factories constructed side by side with worker tenements, commercial businesses, and rowhouses. Industrial pollution made these hybrid neighborhoods dirty and dangerous. Tuberculosis and cholera were rampant, and the infant mortality rate soared.

Early American towns and cities had employed "nuisance laws" to govern the placement and operation of certain harmful or unpleasant industries such as slaughterhouses, stills, and tanneries. But these laws were no longer sufficient for protecting the health and "pursuit of happiness" for urban dwellers, and design ordinances that preserved the aesthetic and character of residential neighborhoods were at this time nonexistent.

Americans were beginning to recognize the need for some limitations on property rights. Concerns over health and overcrowding were part of the reason, but a bigger influence was the negative impact on property values created by this unchecked growth. As industry and the low-rent slums of factory workers began to take over affluent neighborhoods, the affluent were shocked by their plummeting property values. This shock pushed local American governments toward the still-novel concept of municipal zoning.[2]

Zoning, according to the writer Sonia Hirt, is the process of separating land "in a particular area into sections, or zones, with different rules governing activities on that land." Hirt traces zoning's origins to late nineteenth-century Germany, where Frankfurt passed a zoning act in 1891, although similar concepts had already appeared in other parts of Europe. In London, for example, the Metropolitan Building Act of 1844 separated buildings into three separate classes, each with their own standards, although there were no restrictions on where they could be built as in the Frankfurt model.

By the 1900s, American cities desperate to gain control over unregulated growth began looking to these European examples. In 1909, an early urban-planning proponent named Benjamin Marsh published a book titled *An Introduction to City Planning* that espoused European zoning and planning concepts. Marsh's recommendations were soon implemented in New York City, where he served as the influential secretary of the city's "Committee of Population." Soon, Marsh enthusiasts such as the famous landscape architect and planner Frederick Law Olmsted began adopting his ideas.

Although some still consider zoning controversial, the idea took hold across America. Today, approximately 95 percent of American communities, including almost all cities, employ some type of zoning to control the development of private and public property. Hirt lists three reasons for zoning's successful implementation in America: (1) zoning laws protect private property rights by (attempting) to limit government's discretion in their application; (2) zoning laws protect the integrity of private property and its economic value in the American tradition of economic individualism; and (3) U.S. zoning laws help preserve private spaces that promote the American tradition of spatial individualism.

One of zoning's strongest appeals to an urbanizing America was the perception that its principles were applied consistently, with minimal variation, across respective districts. To some degree, this perception of uniformity helped safeguard zoning ordinances from private property laws because they were applied across economic classes, theoretically offering the wealthy and poor equal protections. The detailed codes of many zoning provisions also provided some measure of protection from corrupt politicians, who were forced to rely on the interpretations and applications of more principled government administrators for development approvals. "Hence, American zoning was established to minimize politically (and personalized) motivated decisions about land use," observes Hirt. "It was an attempt to free cities from corrupt dealings and to create clean government, staffed by experts in technical analysis."[3]

However, American zoning has never been immune to these corruptions, and over the decades, this perception of objectivity deteriorated. Large numbers of variances and rezonings are now common, as the American zoning system has become increasingly discretionary over time. This problem is exacerbated in rural, farming areas, where developers take advantage of fragmented community resistance to obtain variances to rezone large tracts for industry or residential housing. And it's no coincidence that both industry and housing subdivisions generate more property tax for local governments than agricultural land and rural homesteads.

These corruptions highlight the fact that zoning has always been an awkward fit for America's farming communities. To standardize zoning practices, many rural communities simply replicated the zoning codes of their urban neighbors, without accounting for the distinct needs and characteristics of farming. Most rezoning or variance processes require some sort of hearing, but rural Americans are often suspicious about participating in public meetings. This problem is exacerbated by the decline of the community newspapers that once served as a common information source for rural areas, where public variance hearings were typically announced. Frequently this leaves the developer, the property owner, and their attorneys as the only "citizen" voices at a zoning variance hearing, with predictable results.[4] But with zoning systems already in place, and generally (if grudgingly) accepted by the majority of American citizens, local officials have turned to zoning concepts for farmland preservation policies, with mixed results.

Agricultural Protection and Cluster Zoning

The American Farmland Trust refers to mandatory county and municipal zoning ordinances designed to protect farmland as "Agricultural Protection Zoning" (APZ) in order to distinguish these ordinances from voluntary "Agricultural District Programs," which are described below. Although no recent effort has been made to catalog agricultural protection zoning ordinances throughout the United States, AFT found seven hundred jurisdictions in twenty-four states with some form of APZs in an informal 1998 poll and includes information on several current examples at its Farmland Information Center.[5]

For instance, the zoning ordinance for Grays Harbor County, located in the state of Washington approximately 50 miles west of Seattle, establishes two designated zones for agricultural use. Parcels zoned "Agricultural Use District" (A-1) attempt to preserve agricultural land for small- to moderate-sized farming activities with a minimum lot size of ten acres. Parcels zoned "Long-Term Agricultural Use District" (A-2) have a minimum lot size of 40 acres and are intended to conserve land for larger, commercial farming operations.

The zoning ordinance of Montgomery County in Maryland creates "Agricultural Reserve" (AR) zones, which restrict residential development to one residential dwelling unit per 25 acres but allows for subdivision of "child lots" of not more than three acres (with exceptions for the feasibility of on-site well and septic systems) for qualified children of landowners. The "Agricultural" (A) zone in the Township of Leacock, Pennsylvania, takes a more fluid approach to minimum lot size requirements, tying subdivisions of A-zoned parcels to the overall size of the "parent" tract from which the subdivision is being made.[6]

Agricultural cluster zoning—also called "conservation development zoning" in some places—is a newer approach to zoning and land development ordinances that attempts to preserve farming areas and prime soils within larger subdivision plans, generally by allowing higher density residential in other areas of the development plan in return for preserving areas with prime soils, or sometimes requiring homes to be sited on areas containing less productive agricultural soils.

However, even a casual reading of these ordinances—rife with exceptions, conditional uses, and variance procedures—reveals the challenges of using zoning as a farmland preservation tool. Cluster zoning is particularly problematic because there has been little empirical study of its overall benefit for preserving commercial agriculture. And what research there is indicates cluster zoning may produce an "Impermanence Syndrome," in which farmers reduce investment in their farming operations as they observe cluster developments proliferating around them, eroding the critical mass of farms and farmland necessary to maintain a viable local agricultural economy.[7] Other critics contend that cluster zoning only facilitates small-scale hobby or "neighborhood" farms—essentially large community gardens—not viable agricultural businesses. Writing back in 1997, even the American Farmland Trust admitted that the function of agricultural protection zoning "is more to limit the pace and density of development than to protect commercial agriculture. In fact, such ordinances often hasten the decline of agriculture by allowing residences to consume far more land than necessary."[8]

A recent literature review, however, provides a more sanguine assessment of agricultural zoning, finding that APZs did reduce the density of subdivision development in designated areas, even if their impacts on future development densities were less conclusive. APZs can also facilitate the development of other farmland preservation mechanisms, such as TDR programs (discussed below), creating an overall additive or synergistic benefit for farmland preservation, the review found. This prompted the study's authors to recommend agricultural zoning as a "primary" farmland preservation tool for local policymakers, although one that should be used in conjunction with other tools such as planning, conservation easements, and PACE programs.[9]

Even with progressive, farmland-friendly zoning in place, many APZs are undermined by state and federal statutes suspicious of local overreach. Ironically, these state interventions often are designed to guard against discriminatory local zoning practices that limit affordable and equitable housing development. State-level interventions "seek to advance economic opportunity, decrease inequality, and further national economic growth by undoing restrictive local zoning," writes Anika Singh Lemar. But local farmland preservation efforts too often are caught in the crosshairs of this state oversight, which is often initiated by the development and home builders' lobbies. Indeed, Lemar's research suggests that state governments provide lobbying opportunities for well-organized interests that cannot compete at the local level due to homeowner and local citizen resistance.[10]

For example, in North Carolina, Chapter 160D of the General Statutes creates a variety of state laws governing the zoning, development, and subdivision of real estate. Article 8 prohibits local regulation of subdivisions greater than 10 acres where no right of way is involved. In theory, this type of regulation may prohibit overzealous local governments from meddling in the private property rights of large landowners with over-restrictive zoning, but in practice it prohibits agricultural-friendly provisions designed to protect and preserve the integrity of farming tracts and prime farmland by maintaining large agricultural tracts.[11]

Farmland Protection and Food System Planning

Another relatively recent development in farmland preservation policy is the increase in farmland protection planning. Often adopting mixed methodologies that include feedback surveys with local citizens, expert or stakeholder interviews, economic impact reports, and some sort of geographic information system (GIS) analysis on current agricultural land inventories and development trends, these farmland protection plans typically identify areas or regions of a planning scope where farming is still viable and farmland is still plentiful, and then offer recommendations for preservation of those areas. These processes can help local governments and community stakeholders set goals for farmland protection, and then identify or propose strategies for their implementation.

In North Carolina, General Statue 106, Article 61, authorizes counties to prepare farmland protection plans that "contain a list of existing challenges to continued family farming in the county" and "describe how the county plans to maintain a viable agricultural community and shall address farmland preservation tools," among other required provisions.[12]

One intriguing part of North Carolina's farmland preservation legislation (nicknamed "Article 61") is that it requires counties to create citizen advisory committees known as "Agricultural Advisory Boards," or AABs, in order to implement and oversee farmland protection planning processes. These boards function similarly to the soil conservation boards envisioned by Hugh Hammond Bennet in the 1930s and '40s by providing opportunities for citizen-led oversight and guidance for local farmland preservation policy.[13] By providing a regular process for citizen engagement and oversight of local farmland preservation, they also create a venue for the "practice of cooperation" that Kemmis describes as such an important element of the place-based solutions and the "politics of place."

Farmland protection plans are increasingly being conducted as part of broader local and regional "foodshed" planning initiatives. "Foodshed" is a term encompassing all stages of food production and consumption in a particular geographic area. Foodshed planning provides recommendations and objectives for maintaining and enhancing these systems, often emphasizing healthy, local food access for socially and economically disadvantaged segments of the community. They typically include proposals not only for farmland preservation but also for the development of agricultural infrastructure—such as local storage, distribution, and marketing resources—that address the broader "agricultural viability" issues necessary for keeping local agricultural communities viable and providing access opportunities for new farmers.

No nationwide survey of farmland protection or food system planning efforts exists today, likely because most of these plans are conducted at the local level, often with the help of third-party contractors like councils of government and private consultants. Perhaps originating in the 1929 book *How Great Cities Are Fed* by Walter Page Hedden, the concept of foodshed planning is now over a century old, but it has re-emerged over the past two decades in response to the social and ecological damage created by the industrialized food system, fueled by interest in the social, environmental, health, and economic benefits of local food. Foodshed planning, write the researchers Megan Horst and Brad Gaolach, "is premised on the unity of place and people of nature and society."[14]

Thus, food systems planning is a policy perspective that emphasizes the interconnected web of cultural, social, economic, and ecological elements within any ecosystem or government realm. These interconnections can be described as "social-ecological systems," or SES, characterizing the inherent dynamism between humans and other living organisms in these systems.

While farmland protection planning is typically focused on the local agricultural community and the farmland conversion problem, often from the land "protection" perspective of resource agrarianism, food system planning typically orients around the triple bottom line objectives of benefiting people, place, and profit. Additionally, food system planning initiatives often call for or adopt the development of "Local Food Councils" to provide citizen oversight and input, with heavy representation from a foodshed's minority and disadvantaged communities, another example of Kemmis's "politics of place" in action.

In their analysis of food systems planning, Horst and Gaoloch suggest these initiatives can be beneficial in evaluating the effectiveness of farmland protection policies such as Agricultural Protection Zoning and PACE programs, which are often recommended within their goals and objectives. These plans also help establish baselines in an agricultural food system for measuring success or implementation of the plans' recommendations. Interestingly, Horst and Gaoloch's analysis found that a major problem in urban areas was that the foodshed in question did not have enough farmland to be food self-sufficient, suggesting that a key element of foodshed planning depends, at least in part, on the existence of viable farmland near urban areas.[15]

Transferable Development Rights and Market Approaches

In a nation where "regulation" can be a dirty word and capitalism is king, some American places attempt to "marketize" farmland preservation through credit-based mitigation systems. An early iteration of these attempts is Transferable Development Rights (TDR) programs, where property owners "sell" their development rights from a designated sending area to a real estate developer who wants to use them in a designated "receiving" area. The landowner who sells these rights subsequently agrees to prohibit or restrict future development on their land through a conservation easement or other form of restrictive covenant. The sending area is usually part of the planning region designated for preservation farmland or other natural resource conservation, while the receiving area is an urbanizing part of the region designated for more intense development. Typically, these development rights are quantified into some form of credits, which are required by local development ordinances.

American TDR programs underwent a wave of popularity in the 1990s and early 2000s. A survey conducted by the independent nonprofit research institution Resources for the Future published in 2007 determined that 140 American jurisdictions (counties, cities, or planning regions) had TDRs on the books. Their analysis did not include an overall estimate of farmland protected through TDRs, but an earlier 2001 study conducted by AFT determined that 67,707 acres of farmland had been protected through TDRs at that time.[16]

One of the reasons there's not more recent TDR research is that, after an initial wave of enthusiasm, their use has generally fallen out of favor within the United States. The Resources for the Future report identified several reasons for this decline. A primary issue was insufficient developer demand for additional density in the receiving areas. "In many communities, trying to force additional density into relatively high-density, already developed receiving areas is very difficult ... In some cases, current residents put up resistance; in others, the demand for higher-density housing is lacking," the report notes.

Another problem highlighted in the report is when development ordinances allow other processes or exceptions for exceeding density limitations within the receiving area—such as density exceptions for connecting to public utilities, cluster housing exceptions, or exemptions for "affordable" housing. These exceptions typically undermine demand for TDR credits as developers learn ways to game the TDR system. And in some cases, TDR requirements are so onerous that developers simply move on to other areas.

On the supply side, the report determined that restrictions associated with the selling of credits in some programs were too severe, discouraging farmer participation and limiting supply. In other programs, the complexity of TDR ordinances discouraged landowner participation, especially when it was unclear upfront how much they would be compensated for selling their TDR rights to a receiver.[17]

However, the report did identify several benefits of TDR programs. When successful, TDRs can generate significant resources for land preservation, without government expenditure. This can be an important benefit when governments don't have other funding sources for conservation easement acquisitions or other protections. TDRs can help compensate landowners for lost value resulting from downzoning; and TDRs can be a good option for communities that want to accommodate growth and preservation, permitting both in appropriate areas of the community. Montgomery County, Maryland, is an example of a place that has successfully employed TDRs in conjunction with zoning and a local PACE program to preserve over 70,000 acres of local farmland. According to the county website, this represents the highest percentage of farmland under agricultural land preservation easements of any county in the nation.[18]

Today, it is more common for local development ordinances that require set-asides or in-lieu fees to support farmland preservation measures. A set-aside is a development ordinance provision that requires a certain amount of land within a development plan to be reserved for agriculture or other open space benefits such as riparian buffers and passive recreation. In Bannock County, Idaho, accessory farm structures such as barns and silos are permitted within its set-aside requirements in an attempt to encourage active agriculture in those areas. An in-lieu fee is essentially a mitigation requirement where the real estate developer pays a fee to offset the development's impacts on agricultural or other open space resources. These fees support the acquisition of farmland or conservation easements in other parts of the community, often through local PACE programs. In Yolo County, California, in-lieu fees are paid to the county's Planning Department, which works closely with the local land trust to identify suitable mitigation properties for permanent protection through conservation easement acquisitions.[19] In-lieu fee programs work similarly to TDR programs without requiring direct, one-to-one preservation in "sending" areas but still creating preservation resources in other areas of the county.

Agricultural District Programs

Although conservation easements and PACE programs create permanent farmland protections, many landowners remain wary of conservation easements' perpetuity and enforcement provisions.[20] Agricultural District Programs, or ADPs, can provide a middle way for these landowners, offering more flexible options for farmland preservation on agricultural properties that aren't a good fit for permanent conservation easements.

ADPs are rooted in early attempts to encourage farmland preservation through reduced property taxes on agricultural property. California's 1965 Land Conservation Act (popularly known as the Williamson Act) was the initial framework for voluntary agreements between local governments and owners of agricultural land. The act authorized local California governments to create "agricultural preserves." Agricultural landowners within these preserves could sign contracts to restrict the use of their land for an annually renewable ten-year period. In return, participants received a reduced property tax rate.[21]

AFT defines ADPs as programs that create special areas or districts where commercial agriculture is encouraged and protected. According to a 2016 AFT report, there are nineteen agricultural district programs in sixteen states, with three states—California, New Jersey, and North Carolina—offering two levels of ADP benefits. In their analysis, AFT found that 63 percent of ADP provisions protect agricultural resources including farmland; 20 percent of ADP provisions attempt to increase agricultural viability on working farmland; and 17 percent of ADP provisions increase security for agricultural operations, such as requiring special hearings for nuisance complaints from neighboring residences.

Enrollment in ADPs is voluntary and typically entitles the landowner to a suite of tax and regulatory benefits. These incentives vary widely across the United States. While most ADPs limit special assessments and the use of eminent domain on participating farms, many ADP programs have unique characteristics. In Wisconsin, ADP participation qualifies landowners for a state income tax credit. In California, ADP participation helps indemnify landowners against legal claims arising from agriculture laborer housing.

In return for these benefits, farmland owners agree to some form of farmland protection for their property, typically in terms limited by years and often with provisions that allow the farm owner to remove the protection provisions without penalty under some situations. These terms also vary widely across states and regions. The Kentucky ADP program requires an initial minimum enrollment of two years, while the Utah program requires a twenty-year commitment. In North Carolina, ADP participants commit to preserve their property for ten years but can withdraw from the program at any time upon providing written notice to the county. Many ADP programs provide landowners with special signage, and some counties hold annual events such as special programming and participant recognition ceremonies for ADP participants.[22]

Conclusion: The Quieter Revolution

The voluntary and incentive-based provisions of ADPs make them an attractive option for local governments and communities seeking more flexible, less restrictive options for farmland preservation than permanent conservation easement. They also create opportunities for social components and place-based governance that fall within the "politics of place" ethos, such as North Carolina's requirement for local agricultural advisory boards. Zoning and planning tools like ADPs, AGZs, TDRs, and food system planning alone cannot stop the escalating loss of American farmland, but they can be effective components of a comprehensive farmland preservation policy.

True, the success of TDR programs and other market-based mechanisms such as in-lieu fees and set-asides is mixed, at best. According to the Resources for the Future TDR analysis, only a few of the 140 TDR programs in the United States it identified during its 2007 analysis were successful at meaningful land preservation.[23] And despite widespread interest about in-lieu fees and set-asides, those programs, where instituted, have also yielded little demonstrable farmland preservation. For instance, as of 2015, the Bannock County set-aside program had not achieved a single on-site easement for agricultural mitigation, while the Yolo County in-lieu fee program for agricultural preservation had been successfully implemented only twice. Both programs were established in 2008.[24]

But if critics complain that these types of farmland preservation tools are ineffectual, largely unmeasured with empirical evidence, and difficult for the public to understand, they are perhaps ignoring their implicit benefits. The planning scholar Robert J. Mason calls these types of planning initiatives a "quieter revolution" that emerged out of the pullback of federal environmental policy in the 1980s and '90s. "Today the emphasis is on place-based, public-private, multistakeholder approaches that treat the human occupants of valued places inclusively and respectfully," he writes. "The quieter revolution is adaptive, attuned, and adjusted to a political climate much more constrained and conservative than that of the 1970s."

According to Mason, elements of this "quieter revolution" include: multistakeholder and collaborative decision making; organizational inclusivity, including land trusts and other environmental advocates but also organizations representing business, educational, and citizen perspectives; local or regional scales of implementation (though frequently guided by federal policy); and some form of associated funding source, either a government ballot initiative or grant resource, private foundation grants, or occasionally punitive awards in response to lawsuits or legal settlements.[25]

And not all success can be measured in numbers. Empirical evidence does suggest that collaborative planning processes can enhance trust among constituents, leading to higher quality decisions, and create a sense of ownership over land preservation processes and outcomes, among other benefits.[26] For instance, the Resources for the Future report argues that, despite their limited efficacy, TDR programs offer an improvement on the status quo, enhancing the overall health, wellness, and natural resource conservation of a community, even if these benefits are, for now, limited. Meanwhile, signage and other marketing tools associated with these programs promote local agricultural and agricultural conservation initiatives, enhancing their farming participants' sense of identity and promoting a farmland preservation ethic and value broadly throughout their local communities.[27]

Ultimately, these findings support a characterization of farmland preservation as a wicked problem, too complex and insidious for any one solution. Critics of these types of "quiet revolution" approaches to farmland preservationists must remember that not all successes are measured in numbers. Even if the individual impacts of these policies so far are limited and inconclusive, they can create cumulative benefits, if not in landscape-scale conservation, then certainly in the way Americans think about farmland preservation and sustainable land use, and the types of government programs and policies they view as acceptable solutions to the problem.

Notes

1 Sonia A. Hirt, *Zoned in the USA: The Origins and Implications of American Land-Use Regulation* (Cornell University Press, 2014), 113.
2 Hirt, *Zoned in the USA*, 9.
3 Hirt, *Zoned in the USA*, 146.
4 Jack White, "New Opportunities for Zoning Reform in Rural Areas," in *Bold Ideas, Emerging Leaders* (Roosevelt Network Emerging Fellows Journal, 2023), 73–82.
5 "Fact Sheet: Agricultural Protection Zoning," American Farmland Trust Farmland Information Center, 1998, https://www.networksnorthwest.org/userfiles/filemanager/1703/.
6 "Local Approaches, Agricultural Protection Zoning, Local APZ Ordinances," American Farmland Trust Farmland Information Center, accessed February 25, 2025, https://farmlandinfo.org/laws/?keywords=agricultural-protection-zoning&level_of_government=local.
7 Thomas L. Daniels, "Where Does Cluster Zoning Fit in Farmland Protection," *Journal of the American Planning Association* 63, no. 1 (1997): 129–137.

8 *Saving American Farmland: What Works* (American Farmland Trust, 1997), 49.
9 David Newburn, Lori Lynch, and Haoluan Wang, "Farmland and Forest Conservation: Evaluation of Smart Growth Policies and Tools," in Gerrit-Jan Knaap, Rebecca Lewis, Arnab Chakraborty, and Katy June-Friesen, Eds., *Handbook on Smart Growth* (Edward Elgar Publishing, 2022), 75–91.
10 Anika Singh Lemar, "The Role of States in Liberalizing Land Use Regulations," *North Carolina Law Review* 97, no. 2 (January 2019), 295–296.
11 North Carolina General Statutes, Chapter 1604, Article 8, https://www.ncleg.gov/EnactedLegislation/Statutes/PDF/ByArticle/Chapter_160D/Article_8.pdf.
12 North Carolina General Statutes, Chapter 106, Article 61, Paragraph 106-744(e), https://www.ncleg.net/enactedlegislation/statutes/html/byarticle/chapter_106/article_61.html.
13 North Carolina General Statutes, Chapter 106, Article 61, Paragraph 106-744(e).
14 Megan Horst and Brad Gaolach, "The Potential of Local Food Systems in America: A Review of Foodshed Analyses," *Renewable Agriculture and Food Systems* 30, no. 5 (2015): 399.
15 Horst and Gaolach, "The Potential of Local Food Systems in America," 404.
16 Robert J. Mason, *Collaborative Land Use Management* (Rowman & Littlefield Publishers, 2007), 158.
17 Margaret Walls and Virginia McConnell, "Transfer of Development Rights in U.S. Communities: Evaluating Program Design, Implementation, and Outcomes," Resources for the Future (2007), 124, https://www.rff.org/publications/reports/transfer-of-development-rights-in-us-communities-evaluating-program-design-implementation-and-outcomes/.
18 "Agricultural Land Preservation," Montgomery County, Maryland, accessed February 28, 2025, https://www.montgomerycountymd.gov/agservices/ag-preservation.html.
19 "Agricultural Mitigation Studies," Land Use & Natural Resources Clinic, University of Montana Law School, Spring 2015, https://farmlandinfo.org/wp-content/uploads/sites/2/2019/09/2015-07-UM-LawClinicAgMitigationCase-Studies-Final.pdf.
20 A.S. DuPont, "Opposition and Barriers to Conservation Easements: Insights from Montana Landowners," MS Thesis, University of Montana, Missoula (2011).
21 N. Bills and R.N. Boisvert, "Information for Evaluating Land Retention Programs: The Agricultural District Approach," Cornell Agricultural Economic Staff Paper No. 88–15 (1988), Cornell University.
22 "Agricultural District Programs," American Farmland Trust Farmland Information Center, 2016, https://farmlandinfo.org/publications/agricultural-district-programs/.
23 Walls and McConnell.
24 "Agricultural Mitigation Studies," Land Use & Natural Resources Clinic.
25 Mason, *Collaborative Land Use Management*, 3–4.
26 Mark S. Reed, "Stakeholder Participation for Environmental Management: A Literature Review," *Biological Conservation* 141, no. 10 (2008): 2417–2431.
27 Andrew Waters, "Agricultural District Programs Provide Flexible Options for Farmland Preservation," *Choices*, Quarter 2, 2024, https://www.choicesmagazine.org/choices-magazine/submitted-articles/agricultural-district-programs-provide-flexible-options-for-farmland-preservation.

Bibliography

American Farmland Trust. 2016. "Agricultural District Programs." https://farmlandinfo.org/publications/agricultural-district-programs/.

American Farmland Trust. 1998. "Fact Sheet: Agricultural Protection Zoning." https://www.networksnorthwest.org/userfiles/filemanager/1703/.

American Farmland Trust. n.d. "Local Approaches, Agricultural Protection Zoning, Local APZ Ordinances." https://farmlandinfo.org/laws/?keywords=agricultural-protection-zoning&level_of_government=local.

American Farmland Trust. 1997. *Saving American Farmland: What Works.*

Bills, N., and R.N. Boisvert. 1988. "Information for Evaluating Land Retention Programs: The Agricultural District Approach." Cornell Agricultural Economic Staff Paper No. 88-15. Cornell University.

Daniels, Thomas L. 1997. "Where Does Cluster Zoning Fit in Farmland Protection?" *Journal of the American Planning Association* 63, no. 1. doi:10.1080/01944369708975729.

DuPont, A.S. 2011. "Opposition and Barriers to Conservation Easements: Insights from Montana Landowners," Master's Thesis. University of Montana, Missoula.

Hirt, Sonia A. 2014. *Zoned in the USA: The Origins and Implications of American Land-Use Regulation.* Cornell University Press.

Horst, Megan, and Brad Gaolach. 2015. "The Potential of Local Food Systems in America: A Review of Foodshed Analyses," *Renewable Agriculture and Food Systems* 30, no. 5. doi:10.1017/s1742170514000271.

Land Use & Natural Resources Clinic, University of Montana Law School. 2015. "Agricultural Mitigation Studies." https://farmlandinfo.org/wp-content/uploads/sites/2/2019/09/2015-07-UM-LawClinicAgMitigationCase-Studies-Final.pdf.

Lemar, Anika Singh. 2019. "The Role of States in Liberalizing Land Use Regulations." *North Carolina Law Review* 97, no. 2.

Mason, Robert J. Mason. 2007. *Collaborative Land Use Management.* Rowman & Littlefield Publishers.

Newburn, David, Lori Lynch, and Haoluan Wang. 2022. "Farmland and Forest Conservation: Evaluation of Smart Growth Policies and Tools." In *Handbook on Smart Growth,* edited by Gerrit-Jan Knaap, Rebecca Lewis, Arnab Chakraborty, and Katy June-Friesen. Edward Elgar Publishing.

Reed, Mark S. 2008. "Stakeholder Participation for Environmental Management: A Literature Review." *Biological Conservation* 141, no. 10. doi:10.1016/j.biocon.2008.07.014.

Walls, Margaret, and Virginia McConnell. 2007. "Transfer of Development Rights in U.S. Communities: Evaluating Program Design, Implementation, and Outcomes," Resources for the Future. https://www.rff.org/publications/reports/transfer-of-development-rights-in-us-communities-evaluating-program-design-implementation-and-outcomes/.

Waters, Andrew. 2024. "Agricultural District Programs Provide Flexible Options for Farmland Preservation," *Choices,* Quarter 2. https://www.choicesmagazine.org/choices-magazine/submitted-articles/agricultural-district-programs-provide-flexible-options-for-farmland-preservation.

White, Jack. 2023. "New Opportunities for Zoning Reform in Rural Areas." In *Bold Ideas, Emerging Leaders.* Roosevelt Network Emerging Fellows Journal.

9 State Policy, Greenbelts, and Open Space Bonds

If America's federal government of the 1970s, '80s, and '90s preferred the neoliberal approach of conservation easements, many of her states could no longer ignore the rapid farmland conversion that was transforming their landscapes. In 1973, Hawaii passed the first statewide farmland preservation legislation in the United States, creating special agricultural districts and zoning designations. Vermont's comprehensive land-use legislation actually preceded Hawaii's in 1970, although their approach focused on planning rather than zoning, and initially did not include specific farmland preservation provisions. Oregon followed suit later in 1973, passing S.B. 100, requiring statewide planning goals for farmland preservation. Soon afterwards, California's 1976 "Coastal Act" passed, mixing Vermont and Oregon's planning approaches with Hawaii's zoning one.[1]

Every state now has some form of farmland preservation policy. But the New Deal's vision for comprehensive land-use reform reflected in Oregon and Hawaii's state programs never gained widespread traction across America, with some notable exceptions. More common in the United States are flexible forms of agricultural zoning, such as those discussed in Chapter 8 and statewide policies that allow local counties to provide economic incentives for farmers.

One type of farmland preservation policy that is ubiquitous across America are state laws permitting differential tax assessments for agricultural lands. Another common approach concerns right-to-farm laws that protect farmers from lawsuits and other nuisance complaints filed by neighbors who are impacted or bothered by farming practices. Greenbelt programs are comprehensive programs that often combine zoning, planning, and acquisition strategies to protect farmland. Although typically formed and administered at the local or regional level, Greenbelts are often an outgrowth of state farmland preservation policy, which create laws and general statutes enabling local governments to adopt more intensive local farmland preservation provisions. Similarly, open space bonds are usually issued by a local or metro region but are enabled through state statutes, which usually require a majority vote in a local public referendum. Open space bonds are government-issued debt certificates used to finance farmland preservation and other forms of open space

DOI: 10.4324/9781003644545-11

protection or recreation infrastructure, such as parks and greenways. Repayment is typically funded through increased property taxes, as approved by voters. Although less common than local initiatives, states can also issue open space bonds for similar objectives. In this chapter we review these examples of farmland preservation programs as we continue our look at current farmland preservation in America.

Differential Tax Assessments

Differential tax assessment programs permit counties to discount the property tax valuation for farm and forestry property ... or other special-use designations. Sometimes called "Present Use Value" programs, "Agricultural Use Value," or similar derivatives, these programs are established by state laws and typically require the landowner to meet some sort of income or management criteria. According to the National Agricultural Law Center, every state in America has legislation authorizing discounted property tax valuations, although the specific provisions and criteria of each law differ widely from state to state.

Differential tax assessments are based on the theory that lowering property taxes on working lands eases the financial pressures some farmers face to sell their land for development and help protect the environmental benefits of large, undeveloped tracts.[2] They acknowledge that undeveloped farm and forestland generally doesn't require the schools, garbage and recycling service, transportation infrastructure, and other county services that can cause property taxes on developed areas to skyrocket. For instance, a Cost of Community Service study conducted for Chatham County North Carolina in 2023 determined that for each one dollar of revenue received by the county for agricultural/forestry uses, the county spent $0.35 to provide required county services. In comparison, Chatham County spent $1.18 to provide county services to residential landowners for each dollar of revenue received, the report determined.[3]

In North Carolina's Present Use Value (PUV) program, three types of property can qualify for differential tax assessment: agricultural, horticultural, and forestland. Qualifying agricultural land must be at least 10 acres of forestland or at least 20 acres in actual production; horticulture requires a 5-acre minimum. The owners of agricultural and horticulture properties must demonstrate to their county governments they have received at least $1,000 in average gross income over the last three preceding years. However, income for property leases, such as leases to harvest hay or graze cattle, do qualify for these income requirements. Since forestland produces income only when timber is harvested, forestland tracts must demonstrate they are complying with a written forestry management plan. The reduction in property tax is calculated by a capitalization rate based on the property's expected net income, which is determined by a Use-Value Advisory Board housed within the North Carolina Department of Revenue.[4]

In Ohio, the differential property tax program is called the Current Agricultural Use Value (CAUV). To qualify, 10 or more acres must be devoted exclusively to agricultural or forestry use; or if under 10 acres, the farm must produce and average yearly gross income of at least $2,500 (alternatively, forestland can be enrolled in the Ohio Forest Tax Law program, or OFTL, which requires that the land is managed according to a certified forestry plan and the landowner attends at least eight hours of forestry education during the first five years of certification). Discount rates for the CAUV program are determined based on the land's soil type, which the state publishes annually.[5]

Given the significant tax savings for properties enrolled in North Carolina's Present Use Value program, most counties audit a percentage of PUV properties annually. Naturally, these audits can be a source of concern for landowners who applied for PUV under false pretenses, or whose original income-producing activities have since ceased. However, North Carolina landowners with a conservation easement on their property are exempt from PUV production or income requirements under certain conditions. To qualify, at least 25 percent of the easement value must be donated by the landowner (i.e., the landowner did not receive financial compensation for that portion of the value lost) and the land must have been enrolled in the PUV program prior to the conservation easement conveyance. Also, the conservation easement cannot restrict commercial agriculture or timber production in any way.

North Carolina land that is disqualified for PUV designation or withdrawn from the program (typically when sold for development) is subject to a three-year "rollback tax" penalty, meaning that the sum of the difference between the PUV designation and the regular property tax rate going back the previous three years must be paid to the county property tax department at the time of ownership transfer.[6] In some North Carolina counties, including Chatham, these rollback taxes are designated toward other farmland preservation efforts, such as PACE programs and conservation easement acquisitions.

Right-to-Farm Statutes

Right-to-farm statutes are state laws meant to protect farmers from nuisance lawsuits. Typically, they apply in cases where the complaining individual or group has established occupation in the area where the farm was previously established or in cases where the farm has existed for a period of time substantially unchanged before the nuisance complaint is filed. According to AFT's Farmland Information Center, "Right-to-farm laws protect farmers who use generally accepted farming practices and conduct reasonable agricultural activities, and also protect them from unreasonable local limits on farming operations."[7]

All fifty states in America have enacted right-to-farm statutes. In a 2019 survey, the National Agricultural Law Center (NALC) created six different categories of right-to-farm laws:

> Triggering Event Statutes: These laws provide additional protections for farm operations based on their precedence over other types of land uses.
>
> Change in Operation Provisions: These statutes describe how changes in farming operations affect the farm's qualification to be considered a nuisance. According to the NALC, only half of American states have these provisions.
>
> Limitations on Protections: These provisions describe the limitations on right-to-farm protections. Generally, they require farming operations to comply with state and federal laws. Some require specific farming practices, defined individually by state.
>
> Preemption of Local Government Actions: These statutes protect qualifying farms from local government actions or ordinances that may conflict with state right-to-farm statutes, or provide additional layers of scrutiny, such as required public hearings on the local enforcement effort.
>
> Attorney's Fees Awarded: These statutes reimburse farmers for attorney's fees and other expenses if the nuisance suit brought against them is deemed to be frivolous or malicious, or if a court finds in favor of the farming operation in a nuisance suit. NALN determined only fifteen states have these provisions.
>
> Damage Caps: These provisions limit the amount of money a farmer is required to pay in compensatory damages in successful nuisance complaints. NALN determined seven states have these protections.[8]

Right-to-farm statutes are popular with farmers, farming advocacy organizations, and agricultural-friendly state legislatures. They create protections for agriculture in urbanizing areas and may provide some sense of security for farmers considering investments in expanding their farm operations. However, a review of case law indicated that these laws often fail to protect the farm in question. One analysis determined that the laws generally favor larger farm operations and may create political pressure for enhancing enforcement of environmental regulations, among other issues.[9] The American Society for the Prevention of Cruelty to Animals (ASPCA) argues that right-to-farm statutes limit the ability of states to regulate the cruel and inhumane treatment of animals and protect other public health threats created by industrial farming activities.[10] Although sometimes controversial, with little empirical evidence supporting their efficacy, right-to-farm statutes are nevertheless one of the most common types of farmland preservation policy in America.

Tax Benefits

Tax benefits in various forms have been a popular strategy for incentivizing farmland preservation and other types of land conservation easements in

America for several decades now. The IRS first affirmed the deductibility of a conservation easement donation in a 1964 Revenue Ruling and codified the benefits for those deductions in the 1976 Tax Reform Act. Today, conservation easements remain the only charitable contributions of a partial interest in property that are ruled deductible for federal income tax purposes.[11]

Congress made the conservation easement deduction provision a permanent part of the Internal Revenue Code (IRC) in 1980, but the specific benefits associated with these donations were established under a series of temporary measures until Congress passed permanent measures in 2015. Under the current law, a landowner making a qualified conservation easement donation can deduct up to 50 percent of the uncompensated conservation easement value from their annual income in any one tax year on their federal tax return. Qualifying farmers and ranchers (generally anyone who receives more than 50 percent of their gross income from "the trade or business of farming") can deduct up to 100 percent of their income. The carry-forward period for this benefit is fifteen years, meaning the donor can take a total of sixteen years (including the tax year of the donation) to "use up" the value of the donated portion of their conservation easement in federal income tax deduction benefits. Significantly, these benefits can apply to a partial donation, meaning that if a landowner is compensated for a portion of the conservation easement (say 75 percent) and donates the remaining portion (25 percent), the donated portion still qualifies for the federal income tax deduction (although the 75 percent that's received in compensation is subject to federal income tax requirements).[12]

Several states now have similar or complementary tax benefits for charitable conservation easement conveyances. In 2022, Land Trust Alliance determined that fourteen states and U.S. territories offer some form of tax credit for conservation easements, and North Carolina has since re-established a tax credit it had previously eliminated in 2013, although only on a temporary basis. Most of these state benefits are in the form of a tax credit, which is a dollar-for-dollar reduction on taxes owed (whereas the federal tax deduction decreases your taxable income). For instance, the state of Georgia allows a 25 percent tax credit on the uncompensated value of the conservation easement donation, although the total tax credit value is capped at $250,000 and must be made to a donee (the organization receiving the easement) that is accredited through the Land Trust Alliance's Accreditation Program. Similarly, in Colorado, qualified conservation easement donations can take a 75 percent tax credit on the first $100,000 of the donation value, then 50 percent on the rest of the value up to $1.5 million. In Florida, a property protected by a permanent conservation easement can receive a 100 percent exclusion from applicable property taxes.

A significant benefit of many (but not all) state conservation tax credit programs is that the credits can be transferred or sold on the open market. For instance, South Carolina conservation easement donors may transfer up to $52,500 in total conservation credits for any single tax year to another taxpayer. These credits are typically sold for some percentage of their overall value. Tax credit transfers can provide additional income opportunities for "land rich, cash poor" landowners who may not otherwise make enough income to be able to fully utilize their tax credits.[13]

Open Space Bonds

Open Space Bonds are a form of government debt issued to raise funds for the acquisition and protection of open spaces—including farmland preservation—and typically fund the development of passive, outdoor recreation projects such as parks and greenways. In essence, open space bonds are not a farmland preservation tool, but they can be used to fund other farmland preservation initiatives such as PACE programs.

Public referendums such as bond initiatives have a long history in North America, with roots in the Populist and Progressive movements of the early twentieth century. These reformers saw public referendums as local tools for promoting issues that lacked support in party-controlled state legislatures, where the lobbyists of corporate interest and the status quo frequently hold sway. Tax and electoral reforms, regulations of railroads, labor rights, and utility regulation were all issues addressed through early American public referenda.[14]

General open space and conservation bond referenda are growing increasingly common in America, fueled by rapid urbanization and growing awareness of its environmental impacts. Often, they are associated with greenbelt planning initiatives such as those described below or some other attempt at landscape-scale open space preservation through comprehensive planning initiatives. The Trust for Public Land (TPL) is an American conservation organization with a long history of facilitating and tracking public referenda for open space preservation and conservation purposes. In a study published in 2010, TPL tracked a total of 2,252 open space bond referenda across forty-five states. Of those, 1,699 passed, securing more than $120 billion for conservation and open space protection.

Open space ballot measures continue to be a popular tool for funding local preservation initiatives. For example, in 2024, the voters of Clay County, Florida, voted to authorize $45 million in bonds for land conservation to include preservation of working forests and farms through fee simple or conservation easement acquisitions. Similarly, Buncombe County, North Carolina, approved a $30 million open space bond referendum in 2022, which also includes funding support for agricultural conservation easement acquisitions.

TPL reports that, of the twenty-three open space funding ballot measures it supported in 2024, all were approved by the voting public, generating over $16 billion in public funding for land conservation and open space preservation, including farmland preservation. However, it is important to note these funding measures, in themselves, do not create farmland preservation, but must be implemented through farmland preservation measures such as PACE programs and other farmland preservation policies and programs discussed throughout this book.

Not all farmland preservation and open space ballot measures are related to bond issuances. For instance, the voters of Jasper County, South Carolina, a rapidly developing county located between the city of Savannah, Georgia, and the popular Hilton Head Island resort area, approved through public referendum in 2024 a 1 percent sales tax increase for a time-limited period of fifteen years to generate an estimated $94 million for open space initiatives, including agricultural conservation easements. In 2020, Chatham County, North Carolina, passed a similar ballot measure increasing the local sales tax, with part of the proceeds subsequently dedicated toward farmland preservation measures.[15]

Open space bond initiatives can occur in both local and statewide contexts, but the laws determining how they can be initiated and used are typically governed by state law and require some type of public approval, either through a stand-alone ballot referendum or as a measure included in a regularly scheduled election. For example, the Chatham County sales tax referendum described above is authorized in Article 46 of Chapter 105 in the North Carolina General Statutes, which stipulates the requirements and administration under which such a public referendum must be conducted.[16] A referendum approving the Article 46 sales tax for Chatham County received a majority in a 2020 vote.

Greenbelts and Urban Growth Boundaries

Greenbelts and urban growth boundaries (UGBs) combine policy, planning, zoning, and public funding initiatives to reduce or prevent urban sprawl into neighboring peri-urban and rural landscapes. Although these urban growth management policies (UGMPS) usually are not specifically focused on farmland preservation, they typically contain some farmland preservation measures, which are often supported through PACE programs and conservation easement incentives. Although implemented at a local level, they require authorizing legislation through the state.

Portland's UGB program is one of the oldest in the United States. It emerged from Oregon's S.B. 100 as an effort to use planning and zoning to preserve farmland and other natural resources against runaway urban sprawl. In theory, an urban growth boundary is a conceptual border around a metropolitan area designed to contain development. When established in 1979, the Portland program used mapping, soil surveys, and other planning tools to identify "exclusive farm use" areas, or EFUs, designated for farmland preservation. Through zoning ordinances, large minimum lots sizes were established within these EFUs, typically 80 acres for farmland and 160 acres for ranchland. EFU designation also limits noncompatible uses on adjacent or nearby parcels, ensuring that "farmers and ranchers can continue to operate by limiting the types and intensity of other uses allowed; provide reduced, or differential, property tax assessments for EFU-zoned parcels; and create a rigorous evaluation process for efforts to approve non-farming development in EFU zones."[17]

Although now modified several times, the Portland UGB has expanded just 14 percent since its 1979 establishment, according to an interview with a Portland planning official. During that time population within the Portland UGB has expanded 60 percent, but 94 percent of growth has occurred within the UGB, suggesting the program is successful at controlling development and preserving farm properties within its boundaries.[18]

The UGB program in Contra Costa County, California, located just west of San Francisco and Oakland, attempts to limit urban development to just 35 percent of the county's land area, preserving the other 65 percent for agriculture, parks, and other open space and conservation purposes through its "Urban Limit Line" program. The measure first passed in a 1990 public referendum and was renewed with some revisions in 2006, with the current measure set to expire in 2026. The areas designated for development and preservation are determined by the county's "Urban Limit Line" map, which is incorporated into the county's open space conservation plan, as well as other county plans, policies, and ordinance codes. The Urban Limit Line was last updated in the 2006 public referendum and can only be changed by approval of four out of the five county supervisors until it expires in 2026.

While not specifically targeted at farmland preservation, the Contra Costa's Urban Limit Line program supports other farmland preservation policies and programs, such as the county's "Agricultural & Open Space Preservation Policy," which establishes strict limits and review procedures for development proposals that may impact the county's agricultural land. In turn, these polices are authorized in California by the state's Cortese-Knox-Hertzberg Local Government Reorganization Act of 2000, which attempts to limit the conversion and loss of California's open space and agricultural lands, among other sustainable initiatives.[19]

A prominent east coast example of a UGMPS policy is Maryland's "Priority Funding Areas" program, which is a part of its broader "Neighborhood Conservation and Smart Growth" legislation initially adopted in 1997. Through this initiative, the state of Maryland requires local governments to prioritize new infrastructure and development projects in already developed areas, with an emphasis on revitalizing older communities. Again, while not specifically a farmland preservation program, this law attempts to preserve farmland by designating specific areas within a county for either development or preservation. It is part of a mosaic of laws and programs that serve as Maryland's farmland preservation policy, including the Maryland Agricultural Land Preservation Foundation, which is a state-sponsored, trustee-governed quasi-governmental authority whose mission is to oversee the preservation of the state's productive farmland and woodland and help curb the spread of "urban blight."[20]

Not surprisingly, many governments tout the successes of their UGB and UGMP initiatives. However, outside researchers have been more measured in their assessments. A 2022 study comparing six United States counties with Greenbelt Programs compared to 19 adjacent counties without greenbelts did find that the greenbelt counties experienced less land conversion between 2006 and 2016, and the rate of population growth in those greenbelt counties was overall lower than the rate of land conversion in the control group counties without greenbelts. However, the study determined that Greenbelt Programs need to be coordinated with other planning efforts addressing affordable housing, adequate infrastructure, and sustainable economic development. "Simply restricting land supply and development activity in a greenbelt could backfire, leading to estate-style housing, leapfrog development, and higher commuting costs," the study's authors' caution.[21]

Similarly, researchers Rebecca Lewis and Robert Parker complain that much of the evidence supporting the effectiveness of UGBs as a tool for farmland preservation has been empirically weak. They worry that the harsh zoning restrictions of UGBs will cause development to leapfrog agricultural areas, isolating these farming areas geographically and ultimately harming the scale and viability of farming there.[22] More disconcertingly, researcher Giulio Giovannoni's 2021 evaluation of Portland's UGB program determined that residential development permitted outside the UGB has reduced the region's overall agricultural productivity. His analysis concluded that urban containment strategies in Portland and elsewhere in Oregon have not been effective in containing urban sprawl.[23]

State Farmland Preservation Programs Highlight: New Jersey and Maryland

While western states like Hawaii, Oregon, and California may have originated state farmland preservation policy in the United States, the concept has flourished in East Coast states like New Jersey and Maryland, where population density and cascading development are exacerbating the farmland conversion crisis. The state of Maryland created the Maryland Agricultural Land Preservation Foundation (MALPF) in 1977, while New Jersey passed its Agriculture and Retention and Development Act (ARDA) in 1983.

Today, the New Jersey Farmland Preservation Program first authorized through ARDA is administered by the State Agriculture Development Committee (SADC), an eleven-member advisory committee that oversees the state's PACE program and other farmland preservation initiative. This committee is supported by staff within the New Jersey Department of Agriculture, who coordinate conservation easements and other program initiatives with local counties and nonprofit conservation organizations.

The SADC's PACE program and the state's other land preservation program are funded through revenues from New Jersey's Corporate Business Tax, a statewide income tax levied on corporations operating within the state. Agricultural landowners who qualify for the PACE program can receive compensation of up to 80 percent of their conservation easement value based on a combination of two appraisals and other defined criterium in a newly adopted "preservation formula." Landowners who agree to additional deed restrictions limiting impervious surface and house sizes can qualify for payment of up to 90 percent of the conservation easement value. Qualifying landowners who apply for the program receive a payment offer based on the scoring criteria, which they can then either accept or reject. As of March 2025, the SADC had acquired over 2,900 farmland preservation conservation easements through its PACE program, according to Stephanie Kreiser, the SADC's chief of acquisition.

According to Kreiser, the preservation formula model was adopted in 2024 due to declining applications for agricultural conservation easements. "We were finding there had been a decrease in applications and people who accepted their offers for preservation over the past twenty years because the value of conservation easements declined by about 20 percent over that period," Kreiser reports. Although the decline was due to a variety of factors, the main reason was that "there just wasn't a big enough difference between the before easement and after-easement value." The state legislature authorized the SADC to test the new formula model for three years to evaluate its effectiveness in increasing farmland preservation applications.

States, municipalities, and qualified nonprofits such as land trusts can also apply for conservation easement funding matched by local funding sources. "For municipal and county partners, we're allowed to cost-share up to 80 percent of the easement value," says Kreiser, although she says a typical cost-share project receives about 60 percent of the conservation easement value from SADC funds, which will also compensate local partners for up to 50 percent of their transaction-related expenses.

Although conservation easement acquisitions are a central part of the SADC's programming, the agency oversees other farmland preservation programs as well, including an agricultural mediation program that can arbitrate disputes between agricultural producers and non-farming neighbors, and a "Farm Link" program that serves as a referral center connecting those seeking farmland and those with farmland available for lease or sale. In our interview, Kreiser highlighted a new "Next Generation Farmer" program that is focused on developing farming opportunities for new and young farmers. "Two of our staff people just spent a year researching the barriers to farming in New Jersey and their report is fascinating. … So the goal is to try to figure out a way to help address these barriers," she said.[24]

In Maryland, state farmland preservation policies are primarily administered through the Maryland Agricultural Land Preservation Program (MALPP), which is governed by the Maryland Agricultural Land Preservation Foundation (MALPF), a thirteen-member board of trustees that includes the state treasurer, secretary of agriculture, and other government executives, as well as nine "at-large" members representing various regions of the state.

Created by Maryland's general assembly back in 1977, the MALPF primarily functions as a PACE program. Most of its funding is provided through a statewide 0.5 percent real estate transfer tax, with a portion of these revenues dedicated to farmland preservation and other land preservation programs. MALPF also receives 25 percent of the revenue from a statewide agricultural transfer tax, collected when agricultural land is sold for nonagricultural development, with certified counties retaining 75 percent of this tax. By leveraging these funding sources, MALPF has purchased over 2,500 agricultural conservation easements, permanently preserving over 350,000 acres across the state.[25]

With support from MALPF and other conservation programs, Washington County, Maryland, has been able to successfully conserve over 12,000 acres of working farmland, despite intensive development along the Interstate 81 and Interstate 70 corridors. "You can reach 80 percent of the population of the United States within a 24-hour drive of Hagerstown," says Washington County's rural preservation administrator Chris Boggs, mentioning the county's largest city and county seat. Due to this location, distribution warehouses are springing up across Washington County's landscape, adding significant revenues to the county's share of the agricultural transfer tax. This boom has allowed Washington County to fully match easement acquisition offers from the MALPF over the last several years, as well as support other local and state preservation programs. "It's the silver lining of the warehouse development boom, if you will," notes Boggs.

One of the unique aspects of Washington County's farmland preservation program is its installment payment plan, which is also funded by the county's share of the real estate transfer tax. According to Boggs, each year the county sets aside $400,000 from its real estate transfer tax revenues to fund the installment plan, available to applicants who "just missed" getting conservation easement funding offers from the MALPP. Landowners who wanted to participate and agree to a conservation easement acquisition price with the county would receive their money in installment payments spread out over ten years. "That way we could preserve as many properties as we could at that time in one year and spread out the payments over the course of ten years," explains Boggs. Participating landowners also receive 2 percent interest on their unpaid balance over the course of the agreement's ten-year payment term.

Despite initial enthusiasm, the county is now moving away from the installment plan to focus on providing match for MALPF easement acquisitions, which provide "the best bang for your buck," according to Boggs. Another prominent farmland preservation tool in Washington County is Maryland's Rural Legacy Program, an initiative of Maryland's Department of Natural Resources that focuses on landscape-scale preservation surrounding the state's most significant natural and cultural assets.[26]

Functioning as both a comprehensive conservation planning tool and a conservation easement acquisition program, the Rural Legacy Program encourages local governments and land trusts to work together to determine how best to protect vital working landscapes through a comprehensive planning process, in essence functioning similarly to a Greenbelt Program. The program is funded through a combination of general obligation bonds from the state's capital budget and dedicated proceeds from Maryland's real estate transfer tax.

In a state that borders the Chesapeake Bay, much of the Rural Legacy Program's land conservation investments are targeted toward its watershed protection. But in Washington County, the program's protections focus on the area around the Antietam Civil War battlefield, sight of one of the bloodiest days in American history. "Our (Rural Legacy Program) concentrates not only on the Antietam Battlefield, but also the approaches and retreats that occurred during the Civil War, so it stretches east to the Appalachian Trail, and then down to where there were approaches and retreats along the Potomac River," explains Boggs.

Applications for the program must come from parcels within the program boundary and are submitted to the state by county programs like Boggs' or through local land trusts, and then evaluated on a combination of factors, including the significance and extent of agricultural, forestry, natural, and cultural resources proposed for protection, along with the threat to resources from development pressure and landscape changes. Projects are also evaluated based on their associated amount of "matching funds" that will also contribute to the conservation easement purchase, originating from local, federal, or private sources.

"It's not a farmland preservation program, per se," explains Boggs. "Part of the way we evaluate these properties is based on agricultural resources, but that's only about a quarter of the points in the ranking system. It also looks at woodland and stream buffers. And the historic nature of the properties is a big part of it." According to the program's website, it has dedicated over $403.5 million to preserve 124,311 acres of farmland, forests, and natural areas since the Rural Legacy Program was established in 1997. Boggs likes the program because it provides more flexible options for landowners. "To me, it's one of our most landowner-friendly programs. It is a little looser when it comes to agritourism and commercial activities. They kind of want the landowners to be able to show off their properties."[27]

Unfortunately, in 2025, Maryland's open space funding was cut by $25 million annually over the next four years, with the funds from the real estate transfer tax redirected to the General Fund to cover other funding priorities. Still, the success of the Maryland and New Jersey farmland preservation programs shows the value of creating dedicated statewide funding streams for farmland and other types of land preservation, building strong local partnerships, and using comprehensive regional planning to achieve significant landscape-scale preservation that ultimately benefits farmland.

Notes

1 Myrl L. Duncan, "Agriculture as a Resource: Statewide Land Use Programs for the Preservation of Farmland," *Ecology Law Quarterly* 14, no. 3 (1987): 401–482.
2 Evin Bachelor and Peggy Kirk Hall, "Differential Tax Assessment of Agricultural Lands," The National Agricultural Law Center, n.d., https://nationalaglawcenter.org/state-compilations/differentialtaxassessment/.
3 "Chatham County North Carolina Cost of Community Services Study," Lois G. Britt Agribusiness Center, University of Mount Olive (December 2023). https://www.chathamcountync.gov/government/departments-programs-i-z/soil-water-conservation-district/farmland-preservation.
4 "Present-Use Value Program Guide," North Carolina Department of Revenue, January 1, 2019, https://www.ncdor.gov/documents/files/present-use-valuation-program-guide/open.
5 "Current Agricultural Use Value General Information," Ohio Department of Taxation, accessed March 2, 2025, https://tax.ohio.gov/government/real-state/cauv.
6 "Present-Use Value Program Guide," North Carolina Department of Revenue, accessed June 25, 2025 https://www.ncdor.gov/present-use-value-program-guide.
7 "Farmland Protection Policies and Programs," American Farmland Trust Farmland Information Center, accessed March 3, 2025, https://farmlandinfo.org/policy-makers-professionals/farmland-protection-policies-and-programs/#state.
8 Alexandra Liazano and Rusty Rumley, "State Right to Farm Provisions," National Agricultural Law Center, 2019, https://nationalaglawcenter.org/state-compilations/right-to-farm-provisions/.
9 Neil D. Hamilton, "Right-to-Farm Laws Reconsidered: Ten Reasons Why Legislative Efforts to Resolve Agricultural Nuisances May Be Ineffective," *Drake Journal of Agricultural Law* 3 (1998): 103–118.
10 "Oppose 'Right to Farm' Legislation," American Society for Prevention of Cruelty to Animals, accessed March 3, 2025, https://www.aspca.org/improving-laws-animals/public-policy/oppose-right-farm-legislation#:~:text=Right%20to%20Farm%20laws%20seek,environmental%20and%20public%20health%20threats.
11 Anson H. Asbury, "Understanding the Conservation Easement Tax Deduction," *The Federal Lawyer*, March 2016, https://www.fedbar.org/wp-content/uploads/2016/03/Easement-pdf-2.pdf.
12 "Using the Conservation Tax Incentive," Land Trust Alliance, accessed March 5, 2025, https://a.storyblok.com/f/120093/x/c6bcecaccd/conservationeasementtaxincentivebrochure2016.pdf.
13 "Land Conservation or Environmental Credits," South Carolina Department of Revenue, accessed March 5, 2025, https://dor.sc.gov/resources-site/lawandpolicy/Documents/SCTIED-2021-Chapter%202-PartF.pdf.
14 Todd Donovan, "Referendums and Initiatives in North America." In Matt Qvortrup, Ed., *Referendums Around the World* (Palgrave Macmillan, 2014), 122–161.
15 "Americans Voted Yes for Water, Wildlife, and Way of Life in 2024," Trust for Public Land, accessed March 8, 2025, https://www.tpl.org/2024-ballot-measures.
16 North Carolina General Statutes, Chapter 105, Article 46, "One-quarter Cent County Sales and Use Tax," accessed March 8, 2025, https://www.ncleg.gov/EnactedLegislation/Statutes/PDF/ByArticle/Chapter_105/Article_46.pdf.
17 "Farmland Protection," Oregon Department of Land Conservation and Development, accessed November 22, 2024, https://www.oregon.gov/lcd/ff/pages/farmland-protection.aspx.
18 Jared L. Green, "Do Urban Growth Boundaries Work to Prevent Sprawl?" Smartcitiesdive, accessed November 22, 2024, https://www.smartcitiesdive.com/ex/sustainablecitiescollective/do-urban-growth-boundaries-work/1070356/.

19 "Agricultural and Open Space Preservation," Contra Costa County, California, accessed March 8, 2025, https://www.contracostalafco.org/policies/agricultural-open-space-preservation/.
20 "Priority Funding Areas," Maryland Department of Planning, https://planning.maryland.gov/pages/OurWork/GDA/pfamap.aspx. Also, "Welcome to the Maryland Agricultural Land Preservation Foundation," Maryland Department of Agriculture, https://mda.maryland.gov/pages/default.aspx. Both accessed May 9, 2025.
21 Albert Tonghoon Han, Thomas L. Daniels, and Chaeri Kim, "Managing Urban Growth in the Wake of Climate Change: Revisiting Greenbelt Policy in the United States," *Land Use Policy* 112 (2022): 11.
22 Rebecca Lewis and Robert Parker, "Exurban Growth inside the Urban Growth Boundary? An Examination of Development in Oregon Cities?" *Growth and Change* 52, no. 2 (2021): 885–908.
23 Giulio Giovannoni, "Urban Containment Planning: Is It Effective? The Case of Portland, Oregon," *Sustainability* 13 (August 2021).
24 Stephanie Kreiser (New Jersey SADC Chief of Acquisition) in discussion with the author, March 31, 2025.
25 "Welcome to the Maryland Agricultural Land Preservation Foundation," Maryland Department of Agriculture, accessed June 28, 2025, https://mda.maryland.gov/malpf/pages/default.aspx.
26 "Maryland's Rural Legacy Program," Maryland Department of Natural Resources Land Acquisition and Planning, accessed June 28, 2025, https://dnr.maryland.gov/land/pages/rurallegacy/home.aspx.
27 Chris Boggs (Washington County, MD, Rural Preservation Administrator) in conversation with the author, March 19, 2025.

Bibliography

American Farmland Trust. n.d. "Farmland Protection Policies and Programs." https://farmlandinfo.org/policy-makers-professionals/farmland-protection-policies-and-programs/#state.

Ansbury, Anson H. 2016. "Understanding the Conservation Easement Tax Deduction," *The Federal Lawyer*. https://www.fedbar.org/wp-content/uploads/2016/03/Easement-pdf-2.pdf.

Bachelor, Evin, and Peggy Kirk Hall. n.d. "Differential Tax Assessment of Agricultural Lands," The National Agricultural Law Center. https://nationalaglawcenter.org/state-compilations/differentialtaxassessment/.

Donovan, Todd. 2014. "Referendums and Initiatives in North America." In *Referendums Around the World*, edited by Matt Qvortrup. Palgrave Macmillan.

Duncan, Myrl L. 1987. "Agriculture as a Resource: Statewide Land Use Programs for the Preservation of Farmland," *Ecology Law Quarterly* 14, no. 3.

Giovannoni, Giulio. 2021. "Urban Containment Planning: Is It Effective? The Case of Portland, Oregon." *Sustainability* 13. doi:10.3390/su132212925.

Hamilton, Neil D. 1998. "Right-to-Farm Laws Reconsidered: Ten Reasons Why Legislative Efforts to Resolve Agricultural Nuisances May Be Ineffective," *Drake Journal of Agricultural Law* 3.

Han, Albert Tonghoon, Thomas L. Daniels, and Chaeri Kim. 2022. "Managing Urban Growth in the Wake of Climate Change: Revisiting Greenbelt Policy in the United States," *Land Use Policy* 112. doi:10.1016/j.landusepol.2021.105867.

Land Trust Alliance. n.d. "Using the Conservation Tax Incentive." https://a.storyblok.com/f/120093/x/c6bcecaccd/conservationeasementtaxincentivebrochure2016.pdf.

Liazano, Alexandra, and Rusty Rumley. 2019. "State Right to Farm Provisions," National Agricultural Law Center. https://nationalaglawcenter.org/state-compilations/right-to-farm-provisions/.

Lewis, Rebecca, and Robert Parker. 2021. "Exurban Growth inside the Urban Growth Boundary? An Examination of Development in Oregon Cities," *Growth and Change* 52, no. 2. doi:10.1111/grow.12481.

10 Farmland Transition

American farmland preservation has advanced significantly in the last fifty years. Conservation easements capitalize on the American spirit of the law by taking agricultural resource protection out of the public government's hands and putting it into those of individual landowners and land trusts. Agricultural District Programs incentivize voluntary preservation through a suite of enhanced benefits and protections for farmers in peri-urban areas. And Greenbelt, Agricultural Protection Zoning, and Transfer of Development Rights programs attempt to incentivize farmland preservation through planning and zoning processes.

But the American farmland preservation movement still struggles to address the farmland conversion that occurs through the inevitable passage of time and the land ownership transition from one farming generation to the next. In an effort to avoid the aristocratic hierarchies of primogeniture—the European tradition of passing real estate to the firstborn child to keep large estates and family wealth intact—the American legal system seemingly encourages the division of assets like prime farmland among heirs and family members. This can make farmland transition feel like the "wickedest" obstacle of all in farmland preservation.

Some argue that a crisis is coming. Inarguably, America's farmers are aging. The United States Department of Agriculture's (USDA) 2022 Census of Agriculture determined that the average age of all farm producers in the United States was 58.1 years, up 0.6 years from the 2017 census. Farmers aged 65 and older comprised 38 percent of all U.S. farm producers, a 12 percent increase from 2017, and farmers aged 65 or older own over 40 percent of American farmland.[1] "So up to 370 million acres of farmland could change hands in the next 20 years, increasing the possibility that the land will be sold for development," warns the American Farmland Trust.[2]

Some resource agrarians dispute these claims. Noting that many farm operations have more than one farm producer, and that the average age of these multi-producer farms tends to be lower than that of farm operations with only one producer, a July 2024 report by the American Enterprise Institute suggests that the farm transition crisis is less dire than organizations like the American Farmland Trust insist. Although midsize and larger farms

DOI: 10.4324/9781003644545-12

represent only 10 percent of all American farms, they produce approximately 80 percent of American agriculture, and these larger farms have substantially younger primary operators than farms classified as small family farm operations, the report argues. These larger farms are also more likely to have formal transition plans in place.[3]

But these resource-based arguments ignore the devastation to rural culture and communities that farmland transition creates in urbanizing areas like Chatham County, where a farm divided up among family members is exponentially more likely to end up in residential development than stay in farming. Certainly, American Farmland Trust's advocacy on the farmland transition issue is a call to action, and their interpretation of the issue should be viewed in that light. Conversely, the American Enterprise Institute is an advocate of private enterprise and limited government, suggesting they are opposed to government policy that engages in a family's land transfer decisions. But these ideological oppositions don't diminish the problem of farmland transition for many American farm families, nor make solutions any easier.

Farmland Inheritance in America: Cotenancy and the "Four *D*s"

For many American farm families, one of the biggest legal challenges they will face is determining an equitable distribution of inherited assets for their children and heirs. Like the Biblical father of the "Prodigal Son," many American farmers want to share their "riches" equally, whether their children have remained on the farm or not. The legal scholars Shannon L. Ferrell and Rodney Jones call this tendency the "Farm Kid/City Kid Conundrum" and describe it as the "Problem of Sweat Equity."

In the "Farm Kid/City Kid Conundrum," one or more children wish to continue the farming operations and possibly continue living on the farm after the senior farming generation has passed. These are the "farm kids." The city kids have no desire to "return" to the farm but still anticipate receiving a fair share of their family inheritance. The senior generation may seek to avoid these conflicts by making all the children co-owners, or cotenants, meaning they each have an equal ownership interest in the farm.[4]

There are three main types of cotenants in the United States: tenancy by the entirety, tenancy in common, and joint tenancy. Tenancy by the entirety is a form of joint ownership specifically for married couples, based in the principle that spouses are an indivisible legal unit. For married couples, tenancy by the entirety provides some important legal protections, such as preventing one spouse from selling or indebting a property without the consent of the other.

In joint tenancy, each cotenant is considered to own the whole estate simultaneously with all the other owners. This ownership share is always equal. For example, if two people own a property as joint tenants, they each own a 50 percent share; for three joint tenants, each owns a 33.3 percent share, and so on. However, each owner has a right of survivorship. This means that upon the death of a co-owner, their interest does not descend to heirs but instead is distributed to the surviving joint tenants. By dying, the tenant loses all claim to the fractional interest, which disappears. The surviving joint tenants stay on a co-owners, each now with a larger fractional interest.

In tenancy in common, each cotenant possesses an undivided interest in the property, meaning each owner possesses the entire property and cannot be excluded from any portion of the property by another owner. However, the ownership interest in tenancy in common need not be the same percentage of overall ownership—in a family of three heirs, for example, one heir can inherit a 50 percent ownership interest while the other two each inherit 25 percent. Nevertheless, a greater percentage of ownership does not necessarily mean a greater influence over management and ownership decisions, unless other legal agreements are made.

Each tenant in common has a separate legal title to their undivided interest, meaning they can mortgage or sell that interest without the consent of the other cotentants. Also, in tenancy in common, a co-owner may pass their ownership interest to their heirs or devisees, who then become tenants in common with the remaining co-owners. There is no right of survivorship, meaning the surviving owners do not acquire the deceased interests unless they are named their heir.

Tenancy in common is the most common form of joint ownership in the United States and exists in all states. Most states have joint tenancy, although some states limit its availability to persons such as trustees and executors of estates. Tenancy by the entirety only exists in 21 states and is not available in community property states, where assets and income acquired during a marriage are equally owned by both spouses.[5]

From a farmland preservation perspective, tenancy in common allows the farmland to remain in an undivided state, preserving its natural resource integrity and agricultural utility. Siblings may initially agree to this type of transition because it honors the parents' wishes to see the farmland preserved within family ownership and gives the farming child the opportunity to continue working the land much as the previous generation did, although with the preferences and considerations of the other siblings as an underlying condition.

But many estate planning professionals are opposed to tenancy in common because of cotenants diverging expectations and financial needs. "This can have a devastating effect on families when heirs have different ideas about what they want done with the assets," writes farmland transition specialist Andrew Branan.[6] This can be especially true in the "Farm Kid/City Kid" conundrum, where the city kid wants to sell the real estate assets and walk away with a cash settlement, while the farm kid wants to continue owning and working the farm.

For cotenants in dispute, partition laws can be manipulated to force a sale, threatening farmland and disrupting the lives of the "farm kid." Partition is a legal process that requires disputing cotenants to divide or sell their interest in the property. In partition by kind, the court orders the land physically divided into smaller parcels that can be separately owned. In partition by sale, the court orders the entire land sold and the proceeds divided among the co-owners.

Both can have devastating effects on the farmland. The threat of partition makes placing farmland in a trust an attractive option for mitigating the potential risks and conflicts of cotenancy. A trust is a legal entity that separates the management and ownership of property from its beneficiaries, who will continue to receive income and benefits from the property. The trust agreement will outline the terms and conditions of these benefits, as well as other terms of the trust, but ultimate authority lies with the trustee—the individual or corporate interest who assumes authority for managing and distributing assets under the trust agreement.

A corporate ownership arrangement, such as a corporation, limited liability corporation (LLC, an American corporate structure that combines the liability protection of a corporation with the tax flexibility of a partnership), or similar legal arrangement can also achieve the long-term management benefits of a trust or be incorporated into the trust agreement. These types of legal agreements do not completely eliminate conflicts between heirs, but they at least specify terms for their resolution. Determining the appropriate legal structure for a particular farm's transition plan is beyond the scope of this book. The point is that relying on cotenant siblings or heirs to agree on farmland preservation over the long term often puts the fate of the farmland itself in jeopardy.

Cotenancy is a fundamental issue in heirs property, a farmland preservation problem that's been gaining recognition in America over the last two decades. Heirs property is land that is jointly owned by descendants of a deceased person whose estate was never handled in probate court, a specialized American court that handles matters related to a person's death and oversees the legalities of the estate administration. This means that heirs property is typically passed down from generation to generation without legal documentation, severely complicating ownership and management issues. Although these joint owners (heirs) retain a legal ownership position in the property, none of them have a clear title. This means that none of these property heirs can sell, mortgage, or enter into legal agreements (such as conservation easements) pertaining to the property because it is unclear who has the legal authority to authorize these agreements. In some cases, speculators can prey on heirs property by convincing one of the heirs to file for partition, then scoop up the property at a reduced price in a forced sale. Unfortunately, many heirs properties are owned by disadvantaged or minority families, increasing the threat from predatory speculators.

The "Problem of Sweat Equity" refers to a situation where one child or heir has been promised a greater share of the inheritance of farmland and farm assets due to their work on the farm. Or they feel as if a greater share is their due because of this work. Often this heir has deferred salary or payment from farm operations in expectation of receiving due compensation when the estate is finally distributed. While this arrangement can be beneficial for farming families, it is too often marred by a lack of communication and changes in the value of the farm assets. Ferrell and Jones note two main problems with the "sweat equity" approach. The main one is insufficient communication. "Rarely is everyone aware of what the arrangement is. Siblings don't know the details of the compensation agreement and are typically shocked when the estate is settled." The second relates to the value of the farm over time. In many cases the value of the farm's real estate assets increases while the value of the farming operation decreases. "Unless one is willing to re-evaluate the estate plan every year, it is extremely difficult to predict the value of the asset base at any point in time and instill any notion of fairness into the final settlement," they note.[7]

Branan characterizes the challenges of farmland transition as the "The *Ds*"—death, disease and disability, and disaster are the dangers that can't be foreseen. Various forms of insurance (life, disability, liability) can be an important tool for mitigating these unexpected dangers. A properly executed power of attorney is another. A greater danger may be the "*Ds*" that result not from the hand of God but from the foibles of human nature: divorce, disagreement, disengagement, and debt. These "*Ds*" force farm families in the farmland transition process to acknowledge the temporality of human relationships and the threat that temporality poses in sound transition planning. "Planning a farm transfer without putting legal tools and arrangements in place to manage the effects of these risks amount to gambling with your future and your family's future," warns Brannan.[8]

Farmland Transition Planning

If conservation easements are the high-profile thoroughbred of farmland preservation, often touted in the media with splashy headlines about acres permanently conserved, farmland transition planning is its workhorse—the tool every single farmland owner absolutely needs. Yet many farmers do it poorly, or they do it in a way that destroys their family legacy of land ownership, or they don't do it at all. Of course, there's no mystery about why this process can be so difficult. It is human nature to avoid the topic of our own death, especially when the discussion reduces us to the sum of our fiscal assets and can lead to uncomfortable conflicts about who gets their share of what.

Branan has defined farmland transition planning as "a process of decision-making that protects your land's agricultural and forest production while preserving family relationships and enhancing community development."[9] And while decision-making, protecting farm and forest, and enhancing community development are all important components of this recipe, preserving family relationships is perhaps its most critical ingredient, the part of the process that so often goes awry.

A qualitative 2009 study by the researcher Matthew Kaplan and his colleagues attempted to quantify the communications issues surrounding farmland transition. Their study identified five main issues influencing family discussions about farmland transition:

Passive communication: Reliance on implicit understanding about respective family roles and transitions regarding farmland estate and transition planning rather than explicit communications.
Delays in planning due to unresolved issues in adult children's lives: Children's long-term career paths and personal relationships remain uncertain long into adulthood, delaying efforts to finalize farm succession plans.
Waiting for children to make career decisions: A child or children's ambivalence about taking over the family farming operations leads to delayed or deferred transition planning.
Concern about the successor's family/marriage stability: Problems in a child's personal relationships prevent the parent from making concrete decisions about formalizing inheritance plans.
Efforts to incorporate children's perspectives into farm management and planning: Generational conflicts over perspectives, values, visions, and concerns delay formal planning.[10]

Certainly, there are other issues that complicate farm transition planning, but this list creates a good understanding about the types of issues that drive the problem. Several recently published research articles focus specifically on communication issues within family farmland transition processes. Communication about succession planning is often complex and uncomfortable, requiring family members to address difficult topics such as retirement, authority, control, and death. Like many families, American farm families tend to make inheritance decisions through tacit understandings and conversations spread out over time and embedded in other activities, making the structure and pressures of planned farm transition discussions stressful for many families.[11]

Ferrell and Jones summarize the common communication issues in farmland transition. One is a "command and control" communication style, where the senior farm operator does not believe there is a need to meaningfully involve successors in planning decisions. Secrecy around transition plans, either from fear of engaging in confrontational discussions or the desire to hide certain aspects of the farm operations from successors, is another. The others are: failure to admit mistakes or fault; unresolved conflicts in other family issues that may influence the transition discussion; and failure to observe "fair" rules for productive discussions and conflict resolutions.[12]

In their research on farmland transition communication issues of Wisconsin farmers, Heather Schlesser and her colleagues focused on the "dialectical tensions" in farmland transition discussion. "I think there's a lot of fear. It's a difficult discussion because you have to talk about your mortality, and you have to talk about the fact of how you can be fair to all your children," one of their study participants reported, observing that it is often easier to delay these discussions rather than address this fear.

Another participant in the Wisconsin study described fears about making and communicating difficult decisions.

> You have to tell them, "This is the way it is. I'm sorry if I hurt your feelings. But if it's not this way, it's not going to work." That's what it comes down to. These farms ... cannot handle splitting them in half, taking a quarter out, paying people off. The margins are too tight to begin with.

Comments such as these prompted the study's authors to recommend third-party mediators for farmland transition planning discussions to help "break the ice and keep the conversation going."[13]

Other research recommends having family members attend group-based educational programming on farmland transition processes and issues as an introduction to the negotiation process, although acknowledging these discussions will ultimately have to happen within their individual families.[14] Researcher Margaret J. Pitts and her colleagues refer to this type of process as dialectical integration, a process of addressing conflicts in transition discussions through the quest for additional information. Meeting with agricultural experts and attending community meetings about farm transfer planning can help families integrate ideas about competing needs and objectives, their research suggests. Sharing information or engaging with difficult topics in separate conversations through segmentation was another strategy their research recommended. "For example, parents might discuss a farming issue before talking about it with their children, who would then share information with their spouses," they write. This "divide-and-conquer" approach can help ease tensions surrounding specific situations by segmenting the topic into several conversations.[15]

If workshops, mediation, and professional advice can help landowners understand the technical process of farmland transition, the more complicated issues surrounding a senior farmer's identity and symbolic capital typically go unaddressed. In an intriguing article from Ireland, researcher Shane Conway and his colleagues studied the role of farmer identity in the farmland transition process. "Farming is not just a job, or something driven by the desire to make money; It is a passion and a lifestyle," they argue, noting that deep-rooted reluctance by elderly farmers to transfer managerial duties to younger generations and "retire" from farming is a significant part of the farm transition issue. "Many farmers' identity and self-esteem are strongly attributed to their occupation and as a result sacrificing one's professional and personal identity upon transferring managerial control of the farm and retiring is a concept they find difficult to accept."[16]

Perhaps not surprisingly, many of the standard approaches to farm transition planning fail to address these emotional issues pertaining to identity and self-esteem. To avoid these complex emotional conflicts, many senior farmers would rather just continue to farm, putting off retirement indefinitely instead of undertaking a meaningful transition planning process. In a survey of a hundred farmers, a 2021 study based in Tennessee found that a large portion of respondents indicated they planned never to retire.[17]

But failure to address the identity dimensions of farmland transfer only leads to delay or avoidance, putting the farmland itself in jeopardy. "It is clear that in spite of the inherent desire to keep the family farm in the family, there is considerable reluctance and resistance among older generations toward relinquishing control and ownership and retiring. Consequently, there is a lack of land mobility (i.e., access to land) leading to an inevitable decrease in the number of farm transfers," observes Conway and his colleagues.[18] The researchers of Tennessee farmers came to a similar conclusion. "While virtually all surveyed farmers felt that farm succession planning is important, as a group they demonstrated both a lack of commitment to and a high level of uncertainty regarding transition planning."[19]

The Wisconsin researchers suggested some novel approaches to addressing identity issues in farmland transition. In their study, twenty-eight Wisconsin farmers participated in focus group discussions about farmland transition. The researchers found that these group discussions produced a cathartic effect for participants. "One participant compared the focus group to a therapy session, as their tensions felt lighter after talking about them in a safe and open environment with peers," the authors write. "We anticipated some reluctance to share experiences. However, participants quickly opened up and shared their very personal stories." Feedback like this led the researchers to recommend peer-to-peer discussion groups as a useful tool in farmland transition programming, noting that several of the participants identified the need for a "shoulder to lean on" when contemplating or entering into difficult family conversations.[20]

American Farmland Transition Policy

Studies like those from Ireland, Tennessee, and Wisconsin cited above only emphasize the psychological complexities of the farmland transition issue. For American farmers, a successful farm transition process can require a variety of professional advisors, ranging from attorneys and certified public accountants, to appraisers, financial advisors, and even family mediators. Adding to the emotional, relational, and psychological stresses of the process are concerns or misinformation about how much these services will cost. And many agricultural landowners struggle to find competent advisors with experience in the unique challenges of farmland transfer.

Given these complexities, it is perhaps no surprise that American policy has struggled to develop meaningful solutions for the farmland transition problem. And it's not as if United States policymakers don't recognize the degree of the problem. The United States Congress authorized the creation of a "Commission on Farm Transitions" in the 2018 Farm Bill. This commission was tasked with studying and developing recommendations on a broad range of farmland transition issues, including the development of federal programs and tax policy to incentivize well-planned agricultural asset transfers and a review of current transition assistance programs, with recommendations for their improvement.

The USDA also provided grant support to the American Farmland Trust to establish the "Land Transfer Navigators" program. The goal of this program was to create a network of trained farmland transfer specialists in partner organizations across the country through trainings, regional forums, and other networking opportunities. American Farmland Trust also offered grants of up to $10,000 to qualifying farm and ranch owners who were actively working with a Land Transfer Navigator organization to assist with farmland transition expenses, including payment for legal and financial planning services.[21]

Sadly, however, Congress never funded or established the Commission on Farm Transition that it authorized in the Farm Bill, and a 2024 bipartisan bill to re-authorize and update the commission passed the House Agriculture Committee but never advanced further in the legislative process. Meanwhile, American Farmland Trust is reorganizing the scope and offerings of its Land Transfer Navigator program after its USDA funding was rescinded by President Donald Trump's incoming administration in 2025, placing its future in doubt.

This leaves the bulk of farmland transition policy to state and local governments, or to the land trust community and local nonprofits. One common approach these organizations try are "Farm Link" programs that attempt to connect new, beginning, or expanding farmers looking for longtime leases and other opportunities to secure farmland with aging landowners who have no heirs or successors interested in managing their farms. According to the American Farmland Trust, there are currently thirty-six Farm Link programs in America, operated by a variety of organizations ranging from academic institutions to state or local government agencies to independent nonprofits.[22] Efforts like these are not solutions to the farmland transition conundrum, but they can at least facilitate or initiate the process.

The Georgia Farm Link Program and the Economics of Inhabitation

Farmland transition is typically portrayed as a family issue, describing the processes and challenges in passing farmland and farm business ownership from one family generation to the next. But not all senior farmland owners have a child or heir who wants to take over the farm. And many senior farmland owners seek reliable and trustworthy tenants for farmland they are not ready to sell but no longer have the capacity to manage. Meanwhile, new farmers struggle to access affordable land.

Farm Link programs link farmland owners seeking tenants with farmers seeking land. Typically, these connections are facilitated through websites and digital message boards, but many Farm Link programs do much more. The North Carolina FarmLink program, for example, offers free, one-on-one farm succession and land access consultations, as well as offering in-person and Internet programs on farmland transition, heirs property issues, and programs for new farmers, among other seminar topics. FarmLink staff also support trainings and programs on farmland transition for other agricultural agencies, such as the North Carolina Cooperative Extension Service and local soil and water conservation districts.[23]

Not all Farm Links are statewide. The Northwest Arkansas Farmlink is a project of the Northwest Arkansas Land Trust that attempts to link farmland seekers with farmland owners specifically in northwest Arkansas's Benton, Carroll, Madison, and Washington counties.[24] The Midwest Organic and Sustainable Education Service, known as Marbleseed, also serves as a Farm Link program. Its listings are not focused geographically, however, but instead toward farm seekers and farmland owners who either practice organic farming systems or want to transition into organic practices.[25] Like the North Carolina FarmLink, both Marbleseed and the Northwest Arksansas Farmlink offer other programs aimed at establishing or maintaining agricultural viability, although most are targeted at beginning or novice farmers, not senior farmland owners.

The Georgia Farm Link program began in 2018 as an initiative of the Athens Land Trust. In the beginning, the program functioned as a traditional matching service, linking farmland owners with farmland seekers through the Internet. According to Nigel Pritchard, the director of the Georgia Farm Link program, the Website listed approximately 260 farm seekers and seventeen available farms on the day we talked in the spring of 2025.

But like the North Carolina Farm Link program, and many others across the United States, the Georgia Farm Link program aspires to be a comprehensive resource for navigating the farm transition process. "At the moment, we're also building an educational site so we can actually do training for succession planning and farmer business development and even starting your own farm and how you go out and start from the very beginning," explains Pritchard.

Initiatives like this reflect a growing interest in programs that help new or small-scale farmers gain access to farmland, not simply approaching farmland preservation as a resource protection problem, but also focusing on agricultural viability by providing new pathways to land access for beginning farmers, and then helping them with financial and business development training so they can build and maintain a viable farm business that keeps the farmland in active agricultural production. Pritchard muses:

> Our vision for Georgia Farm Link is to help a beginning farmer become an incubator farmer, then eventually find their way to land through Farmlink, eventually purchase that land because of our ability to get them ready to have a farm business, eventually put it into a conservation easement to protect it, then pass it on to their kids.[26]

Although they are often operated on a statewide scale, Farm Link programs harness the power of the Internet to connect farmland owners with farmland seekers on a local level. True, these connections are not always equal—like Georgia Farm Link, most Farm Link websites seem to profile more farmland seekers than offerors. Still, their spirit functions within Kemmis's vision for the politics of place and its emphasis on building local markets. "It may well be that neither towns nor farms can thrive in the way they would prefer until they turn their attention more directly to each other," Kemmis observes.[27] And although he is referring in this passage to a more traditional definition of market—the one between farm producer and local buyer—the Farm Link concept is based on the same principle, except that it marketizes available farmland in a way that sustains and grows the local agricultural community.

Farm Link programs also embrace the iteration of the "politics of place" endorsed by scholar A.S. Cheng and his publishing colleagues, who define it as collaboratives of stakeholders from diverse backgrounds organized around place-based connections to address common resource management issues. "The common thread running through these initiatives is the notions that interactions occurring in place-based collaborations tend to center on problem solving, emphasize trust building, and focus on achieving on the ground actions," they write.[28]

The market for Farm Links may be statewide or regional in scale, but the connections they create are local, forging connections in place. And the networks they produce are based in a shared ethic of farmland preservation. If this is not a perfect example of Leopold's "conservation ethic," which puts the health of the land and the ecosystem above the economic needs of humans, it is at least one that marketizes the kind of stewardship ethic for both humans and land that a "New Agrarian" would find satisfactory.

And Kemmis appears to have more faith in the markets of place than its politics. Speaking of the various benefits that come from place-based collaborations, he writes, "as a rule, we come closer to this way of thinking in the economic rather than in the political sphere." In one part of *Community and the Politics of Place*, he calls this process the "economics of inhabitation ... the finite bounded setting within which the age-old interplay of scarcity and possibility, which we call the economy, may unfold itself."[29]

Pritchard may not be specifically referring to the economics of inhabitation when he speaks of his cradle-to-the-grave vision for Georgia Farm Link—providing service and support for the entire cycle of the farmland transition process, from training new farmers and helping them access land to sustainable farm management to transitioning farmland ownership to the next generation—but the goal is the same: a new model of American agriculture that grows and nurtures farming communities through supporting new farmers while it values and preserves farmland. Through this spirit of stewardship, Farm Link programs can play an important role in farmland preservation and the politics of place, which is considered in the next section.

Notes

1 "Highlights from the 2022 Census of Agriculture: Farm Producers," United States Department of Agriculture, February 2024, https://www.nass.usda.gov/Publications/Highlights/2024/Census22_HL_FarmProducers_FINAL.pdf.
2 "Farms Under Threat 2040: Choosing an Abundant Future," American Farmland Trust, accessed August 9, 2024, https://farmland.org/project/farms-under-threat/.
3 Eric J. Belasco and Joseph W. Glauber, "An Aging Farm Population: Cause for Concern?" American Enterprise Institute, July 15, 2024, https://www.aei.org/research-products/report/an-aging-farm-population-cause-for-concern/.
4 Shannon L. Ferrell and Rodney Jones, "Legal Issues Affecting Farm Transition" (paper presented at the Transitions in Agriculture: Implications for Research, Data Development and Policy Analysis Conference, Washington, D.C., March 20–21, 2013).
5 J. David Reitzel, Robert B. Bennett, and Michael J. Garrison, *American Law of Real Estate* (Thomson Learning, 2002), 125–129.
6 Andrew Branan, "Planning the Future of Your Farm: A Workbook on Farm Succession," Green Fields Sandhills and the North Carolina Cooperative Extension Service, n.d., 13.
7 Ferrell and Jones, "Legal Issues," 15.
8 Branan, "Planning the Future of Your Farm," 13.
9 Branan, "Planning the Future of Your Farm," 10.
10 Matthew S. Kaplan, Jon F. Nussbaum, Craig Fowler, and Margaret J. Pitts, "Communication Barriers to Family Farm Succession Planning," *Journal of Extension* 47, no. 5 (Oct. 2009).
11 Margaret J. Pitts, Craig Fowler, Matthew S. Kaplan, Jon Nussbaum, and John C. Becker, "Dialectical Tensions Underpinning Family Farm Succession Planning," *Journal of Applied Communication Research* 37, no. 1 (2009): 59–79.
12 Ferrell and Jones, "Legal Issues."
13 Heather Schlesser, Sandra Stuttgen, Liz Binversie, and Joy Kirkpatrick, "Insights into Barriers and Educational Needs for Farm Succession Programming," *Journal of Extension* 59, no. 4 (2021): 2–3.
14 Megan L. Roberts, "Farm Management Instructors Expect More Farm Transitions in Coming Years," *Journal of ASFMRA* (2021): 29–34.
15 Pitts et al., "Dialectical Tensions," 73–74.
16 Shane Conway, John McDonagh, Maura Farrell, and Anne Kinsella, "Cease Agricultural Activity Forever? Understanding the Importance of Symbolic Capital," *Journal of Rural Studies* 44 (2016): 4–5, 29.
17 David W. Hughes, Robert Holland, Joey E. Mehlhorn, and Anthony Delmond, "Evaluating Transition Planning by Tennessee Farmers," University of Tennessee Extension Institute of Agriculture (2019).
18 Conway et al., "Cease Agricultural Activity Forever?" 12.
19 Hughes et al., "Evaluating Transition Planning," 6.
20 Schlesser et al., "Insights into Barriers and Educational Needs," 8.
21 "Farm and Ranch Transition Support Grants," American Farmland Trust, accessed July 2, 2025, https://aft.smapply.io/prog/farm_and_ranch_transition_support_grant/.
22 "Nationwide Farm Link Finder," American Farmland Trust Farmland Information Center, accessed July 2, 2015, https://farmlandinfo.org/farm-link-finder/.
23 "NC Farmlink," North Carolina State Cooperative Extension, accessed on October 24, 2025, https://ncfarmlink.ces.ncsu.edu/.
24 "Connecting Farmers and Farmland," NWALT Farmlink, accessed October 25, 2025, https://nwafarmlink.org/.
25 "About Marbleseed," Midwest Organic and Sustainable Education Service, accessed October 25, 2025, https://marbleseed.org/programs.

26 Nigel Pritchard (Director of Georgia Farmlink) in discussion with the author, April 8, 2025.
27 Daniel Kemmis, *Community and the Politics of Place* (University of Oklahoma Press, 1990), 125.
28 A.S. Cheng, Linda E. Kruger, and Steven E. Daniels, "Place as an Integrating Concept in Natural Resource Politics: Proposition for a Social Science Research Agenda," *Society and Natural Resources* 16, no. 2 (2003): 88.
29 Kemmis, *Community and the Politics of Place*, 102.

Bibliography

American Farmland Trust. "Farms Under Threat: Choosing an Abundant Future." Accessed August 9, 2024. https://farmland.org/project/farms-under-threat.

American Farmland Trust Farmland Information Center. "Farm and Ranch Transition Support Grants." Accessed July 2, 2025. https://aft.smapply.io/prog/farm_and_ranch_transition_support_grant/.

American Farmland Trust Farmland Information Center. "Nationwide Farm Link Finder." Accessed July 2, 2025. https://farmlandinfo.org/farm-link-finder/.

Belasco, Eric, and Joseph Glauber. 2024. "An Aging Farm Population: Cause for Concern?" American Enterprise Institute. https://www.aei.org/research-products/report/an-aging-farm-population-cause-for-concern/.

Branan, Andrew. n.d. "Planning the Future of Your Farm: A Workbook on Farm Succession," Green Fields Sandhills and the North Carolina Cooperative Extension Service.

Cheng, A.S., Linda E. Kruger, and Steven E. Daniels. 2003. "Place as an Integrating Concept in Natural Resource Politics: Proposition for a Social Science Research Agenda," *Society and Natural Resources* 16, no. 2. doi:10.1080/08941920309199.

Conway, Shane, John McDonagh, Maura Farrell, and Anne Kinsella. 2016. "Cease Agricultural Activity Forever? Understanding the Importance of Symbolic Capital," *Journal of Rural Studies* 44. doi:10.1016/j.jrurstud.2016.01.016.

Ferrell, Shannon L., and Rodney Jones. "Legal Issues Affecting Farm Transition," paper presentation at the Transitions in Agriculture: Implications for Research, Data Development and Policy Analysis Conference, Washington, D.C., March 20–21, 2013.

Hughes, David W., Robert Holland, Joey E. Mehlhorn, and Anthony Delmond. 2019. "Evaluating Transition Planning by Tennessee Farmers," University of Tennessee Extension Institute of Agriculture.

Kaplan, Matthew S., Jon F. Nussbaum, Craig Fowler, and Margaret J. Pitts. 2009. "Communication Barriers to Family Farm Succession Planning," *Journal of Extension* 47, no. 5.

Kemmis, Daniel. 1990. *Community and the Politics of Place*. University of Oklahoma Press.

Midwest Organic and Sustainable Education Service. "About Marbleseed." Accessed October 25, 2025. https://marbleseed.org/programs.

North Carolina FarmLink. "About NC FarmLink." Accessed October 24, 2025. https://ncfarmlink.ces.ncsu.edu/.

Northwest Arkansas Farmlink. "Connecting Farmers and Farmland." Accessed October 25, 2025. https://nwafarmlink.org/.

Pitts, Margaret J., Craig Fowler, Matthew S. Kaplan, Jon Nussbaum, and John C. Becker. 2009. "Dialectical Tensions Underpinning Family Farm Succession

Planning," *Journal of Applied Communication Research* 37, no. 1. doi:10.1080/00909880802592631.

Reitzel, J. David, Robert B. Bennett, and Michael J. Garrison. 2002. *American Law of Real Estate*. Thomson Learning.

Roberts, Megan L. 2021. "Farm Management Instructors Expect More Farm Transitions in Coming Years," *Journal of ASFMRA*: 29–34.

Schlesser, Heather, Sandra Stuttgen, Liz Binversie, and Joy Kirkpatrick. 2021. "Insights into Barriers and Educational Needs for Farm Succession Programming," *Journal of Extension* 59, no. 4. doi:10.34068/joe.59.04.04.

United States Department of Agriculture. 2024. "Highlights from the 2022 Census of Agriculture." https://www.nass.usda.gov/Publications/Highlights/2024/Census22_HL_FarmProducers_FINAL.pdf.

Part Three

American Farmland Preservation and the Politics of Place

11 Farmland Preservation, Social Capital, and Sense of Place

Part of my job as farmland preservation coordinator is working with our Agricultural Advisory Board. This is a committee of citizen volunteers who meet every three months to provide citizen oversight and guidance over various aspects of our farmland preservation program. It also works to make sure the issues of our farming community are communicated throughout the county government.

The Agricultural Advisory Board is one of many similar citizen committees and boards functioning within our county government. Others provide similar citizen input and guidance over issues such as environmental sustainability and efficient energy use, affordable housing, and economic development, to name a few.

Perhaps you've served on such a committee for your own community. And perhaps like many, you've had a mixed experience there, finding the participation engaging and meaningful at times, pointless and mindlessly bureaucratic at others. I suspect the members of my Agricultural Advisory Board feel the same and, in truth, the body is bureaucratic in origin. In North Carolina, agricultural advisory boards are required by the North Carolina General Statutes, at least for counties that want to participate in various state-sanctioned farmland preservation programs, such as our Voluntary Agricultural District program, or perform other actions related to farmland preservation authorized within state law.[1]

But outside of these bureaucratic, administrative, and legal responsibilities, the Agricultural Advisory Board creates social benefits, both for its members and the broader agricultural community of Chatham County, even if these benefits often seem buried deep beneath its legislated roles. Sometimes we get together at members' houses or places of business. Sometimes we eat together. Often, we talk informally about farms and farming issues in the county outside of our officially designated roles. And sometimes we meet with county representatives from other oversight committees, or county staff from other departments, or even the county commissioners themselves, and share our discussions with them.

In sociological terms, the social byproduct of this civic engagement process can be defined as "social capital," the social and cultural benefits

DOI: 10.4324/9781003644545-14

created through social interactions championed by sociologists such as Pierre Bourdieu and Robert Putnam. In his famous book on American society, *Bowling Alone*, Putnam argued that one of the main problems in American society is the decline of social capital. Sometime around the end of World War II, he believed, Americans became less involved in the civic organizations and community groups that were, up to that time, a hallmark of American society. Television, suburban sprawl, families with two working parents, and a general shift toward a more individualistic culture all shared the blame for this societal shift in Putnam's analysis.

These influences, and no doubt others (originally published in 2000, *Bowling Alone* came out at the beginning of the Internet age, so it doesn't consider the societal impacts of digital culture), created an overall decline in social capital, contributing subsequently to a variety of social issues. One of the biggest impacts of this societal shift, Putnam argues, is the decline of what he calls civic virtue, or regular participation in democratic process. "The core idea of social capital theory is that social networks have value," Putnam explains.

> The difference is that "social capital" calls attention to fact that civic virtue is most powerful when embedded in a dense network of reciprocal social relations. A society of many virtuous but isolated individuals is not necessarily rich in social capital.

Putnam's ideas on social capital were heavily influenced by sociologist James Coleman. An education researcher, Coleman famously used data from a large national sample of public school students in the United States to link supportive social networks with higher high school graduation rates, concluding that social capital was instrumental in efforts to transcend disadvantaged socioeconomic backgrounds.

Coleman framed his ideas on social capital through rational choice theory, arguing that human action is primarily motivated by self-interest. For Coleman, cooperation among actors pursuing their own self-interests accrues not only to the individual but also to others within their social groups. This process of cooperative accumulation provides individuals in pursuit of individual goals access to otherwise unavailable resources.

Coleman's depiction of social capital as a byproduct of cooperative accumulation led him to focus on the role of social capital within organizations. Coleman defined "appropriable social organizations" as ones formed for one purpose that can be appropriated for others. For example, the social capital accumulated within an organization formed to bring attention to the decrepit condition of a local playground can later be accessed for a broader park planning initiative. In contrast, an "intentional organization" is one that is organized to directly benefit those who are invested in its formation but can also benefit those less immediately involved. A political action committee organized to elect a specific candidate could meet this description.

Presumably, the people who organize and contribute to this committee expect some direct benefit from the candidate's election, although others who share the candidate's political values could also benefit if the candidate is elected.[2]

Coleman's civic and group-oriented characterization of social capital is often contrasted with the resource-based social capital theories of the French sociologist Pierre Bourdieu, who focused on social capital's role in class distinctions and social power. Bourdieu defined social capital specifically as "the sum of resources, actual or virtual, that accrue to an individual or a group by virtue of possessing a durable network of more or less institutionalized relationships of mutual acquaintance and recognition."[3] For Bourdieu, social capital is not an individual possession but a collective possession of the group employed in the pursuit of social power and preserving group interests. Social capital, then, allows the members of the group to access, manipulate, and maintain class and group interests. Social capital can also be deliberately created through the cultivation of social ties. He famously described social capital in terms of membership in a golf club, which he described as helpful in accessing social power and status for its members.[4]

Although Putnam acknowledged Bourdieu's influence, *Bowling Alone* was primarily focused on Coleman's "civic" iteration, using social capital as a theoretical lens for America's declining civic engagement. Putnam chose the image of the solo bowler to symbolize this decline, emphasizing the way the simple act of bowling alone reverberates through local communities and economies. At the time, the number of Americans participating in league bowling had declined 40 percent since its peak in the 1950s and '60s. He wrote:

> The decline of league bowling threatens the livelihood of bowling lane proprietors, because according to the owner of one of the nation's largest bowling lane chains, league bowlers consume three times as much beer and pizza as do solo bowlers, and the money in bowling is in the beer and pizza. The broader social significance, however, lies in the social interaction and even occasionally civic conversations over beer and pizza that solo bowlers forgo.[5]

There's a good reason Putnam didn't call his book *Farming Alone* or depict a solo farmer on its cover. Farming has always been portrayed as a solo enterprise in American culture, epitomized by the psychological isolation captured by artists such as Grant Wood in his masterpiece, *American Gothic*, or the famous Dust Bowl photography of Dorothea Lange.

But community and civic participation were important components of early American farming. In his work on English folkways titled *Albion's Seed*, historian David Hackett Fisher described the community settlement patterns of early English settlers in Virginia's Chesapeake Bay region:

> The primary units of belonging were the family and the rural "neighborhood," a word often used in the seventeenth or eighteenth centuries to describe a group of ten or twenty households which were close enough to think of themselves as neighbors.

As generations of these farming neighbors married, they created kin groups tied by relationships of increasing density. The members of these kin-neighborhoods formed closed social networks that worked and socialized together. They even borrowed from one another, adding economic ties to their social ones. And the scale of community in these early farming neighborhoods expanded broadly on church and court days. Then, "Virginians of all ranks and conditions met and mingled at their parish churches and county courthouses, which were favorite places for buying and selling, racing and gambling, meeting and gossiping," writes Fisher.[6]

The integrity of these rural kin groups declined in the twentieth century, but they remained an important symbol of rural community. And their social capital could still be accessed in times of need or cultural significance. In *Community and the Politics of Place*, Kemmis recalls a "barn-raising" party on his grandfather's Montana farm. "It never would have occurred to us, in the early 1950s, to tackle this massive job without calling on the neighbors for help," Kemmis recalls, pointing to the memory as an example of the social capital that still existed in the rural communities of his youth.

Kemmis frames this particular memory through the story of a neighbor named Albert, whose libertine lifestyle was not approved of by Kemmis's mother, Lilly.

> In another time and place, Albert and Lilly would have had nothing to do with one another. But on those Montana plains, life was still harsh enough that they had no choice. Avoiding people you did not like was not an option. Everyone was needed by everyone else in one capacity or another.[7]

Kemmis's memories evoke Coleman's theory of cooperative accumulation, suggesting individual goals spur group interactions. The nineteenth-century French author and sociologist Alexis de Tocqueville also observed this American proclivity toward collective action motivated by individual self-interest. In his landmark work, *Democracy in America*, published in two volumes in 1835 and 1840, Tocqueville identified America's zeal for association as a defining characteristic of its democratic principles. "Americans of all ages, all conditions, all minds constantly unite," he marvels. "As soon as several of the inhabitants of the United States have conceived a sentiment or an idea that they want to produce in the world, they seek each other out." And through this instinct to come together, Tocqueville observed, we overcome the dehumanizing effects of democratic society. After association, Tocqueville argues, Americans "are no longer isolated men, but a power one sees from afar … a power that speaks, and to which one listens."[8]

Indeed, Tocqueville believed civic participation, which he termed "association," was the very essence of American democracy.

> There is scarcely an undertaking so small that Americans do not unite for it … Thus, the most democratic country on earth is found to be, above all, the one where men … have most perfected the art of pursuing the object of their common desires in common.[9]

Some might argue Tocqueville foresaw the decline in civic participation Putnam laments in *Bowling Alone*, anticipating in it disastrous consequences for American democracy. "The morality and intelligence of a democratic people would risk no fewer dangers than its business and its industry if the government came to take the place of associations everywhere," Tocqueville wrote. He predicted that in the absences of association, government "will exercise an insupportable tyranny even without wishing to; for a government knows only how to dictate precise rules … It is therefore necessary that it not act alone."[10]

If it is perhaps difficult to foresee the collapse of American democracy warned of by Tocqueville in the decline of the bowling league, Putnam argues convincingly that re-establishing the social norms embedded in this intrinsic type of civic participation could go a long way toward ameliorating our current era of political and social divisiveness. "Frequent interaction among a diverse set of people tends to produce a norm of generalized reciprocity. Civic engagement and social capital entail mutual obligation and responsibility for action," Putnam writes. "When economic and political dealing is embedded in dense networks of social interaction, incentives for opportunism and malfeasance are decreased."[11]

That is a lot of responsibility to put on a bowling league, and one of the criticisms of *Bowling Alone* is that it ignores the role of self-interest in civic participation described by Coleman and Bourdieu. People don't join a bowling league to create social capital, critics argue; they do it to have fun, make friends, perhaps meet a prospective mate, and maybe generate a business lead or two. In response to those criticisms, Putnam points toward Tocqueville, whose doctrine of "Self-Interest Well Understood" posited that early Americans were willing to sacrifice some of their self-interest because they understood such sacrifice enhanced the preservation of others. In short, the bowlers of the 1950s and '60s joined leagues to have fun *and* create collective community benefits, recognizing the "win-win" scope of social capital. If Americans are not by nature virtuous, inherent in our values is an understanding that virtue is nevertheless useful and can often be rewarding.[12]

Other criticisms of *Bowling Alone* argue that Putnam's research overlooks the complexities of race, power, and privilege embedded in American society.[13] But even acknowledging its weaknesses, *Bowling Alone*'s influence in popularizing the social capital theories of Bourdieu, Putnam, and others is impossible to ignore. One of these influences is Putnam's use of the terms "bonding social capital" and "bridging social capital" (frequently shortened to just "bonding capital" or "bridging capital") to differentiate the ways that social capital functions within group or community dynamics. Bonding capital refers to the social ties within homogenous groups that tend to reinforce exclusive identities, such as the social capital that exists within the same family, club, or small church. County commissions and city councils can also be dominated by bonding capital, particularly if the membership in these governing bodies is static and made up of people from similar backgrounds. In these smaller, more confined social networks developed over time, bonding capital enhances trust and agency, and helps intragroup members navigate conflicts.

Bridging capital, in contrast, develops through diverse, heterogenous networks that are outward-looking and encompass people across social divides and backgrounds. "Bridging capital can generate broader identities and reciprocity, whereas bonding capital bolsters our narrower selves," describes Putnam. "Bonding capital constitutes a kind of sociological superglue, whereas bridging capital social capital provides a sociological WD-40."[14]

In his research on natural resource governance, the scholar Carl Folke and his colleagues found that bridging relationships helps link local actors and communities with higher scales of government, mitigating the divisions between these levels that often plague traditional, hierarchical government approaches.[15] Other bridging capital research suggests a positive collaborative effect when participants from different professional and personal backgrounds are connected in a governance process. Members of these "bridging" social networks tend to develop more trust in collaborative government and are more confident in their ability to solve conflicts.[16] Bridging capital can also lower the cost of collaboration and conflict resolution, facilitate trust and coalition-building, provide access to government resources and capital without limiting entrepreneurial initiative and innovation, and promote adaptive co-management of local natural resources.[17]

Given research findings like these, one might suspect that bridging capital is the main ingredient in the type of social capital that facilitates civic virtue. Indeed, bonding capital's inward-looking focus is sometimes depicted as the "dark side" of social capital. Some research indicates homogenous networks built from tight local bonds can create inertia-based, normative social environments, entrenching traditional power structures, excluding minority segments of the community, and restricting the ability to respond to social and environmental crises through change and innovation.[18]

But both bonding and bridging social capital play vital roles, and both are necessary for civic participation. The solidarity formed through bonding capital may be especially important in communities facing outside threats, such as growth or overdevelopment, or where narrative of shared backgrounds and values are necessary to overcome entrenched political resistance.[19] "Under many circumstances both bridging and bonding capital can have powerfully positive social effects," agrees Putnam, who observes that many groups "bond" under some social conditions and "bridge" under others. "In short, bonding and bridging are not 'either-or' categories into which social networks can be neatly divided, but 'more or less' dimensions along which we can compare different forms of social capital."[20]

Sense of Place

Both bridging and bonding social capital both play a role in forming the social characteristics of place. "It may well be that the most meaningful places, those that are the objects of the most intense attachments, are those which have a potential for combining both bridging and bonding capital," writes the social science researcher Maria Lewicka, "and that the mediating factor is the ability of such places to facilitate both types of social capital, instead of only one."[21]

Lewicka's research is framed through the theoretical paradigm of "sense of place." The early place theorist Yi-Fu Tuan defines sense of place (SOP) as the meanings and attachment to a setting held by an individual or group.[22] Since Tuan's groundbreaking work, sense of place theory has evolved into a true multidisciplinary field with growing bodies of research in areas ranging from architecture, planning, and urban design to environmental psychology, environmental philosophy, geography, public land managements, and leisure and recreation.

As Tuan noted, "place exists at different scales," ranging from country (or even the planet in the growing awareness about global warming) to the neighborhood. Tuan defined the "homeland," as the "region (city or countryside) large enough to support a people's livelihood," and noted that "Attachment to the homeland can be intense … Human groups nearly everywhere tend to regard their own homeland as the center of the world."[23]

Scholar J.B. Jackson contends that sense of place is based on ritual repetitions within community or place contexts, what he calls "a lively awareness of the familiar environment" and a

> sense of fellowship based on shared experience … What brings us together with people is not that we live near each other, but that we share the same timetable: the same work hours, the same religious observances, the same habits and customs … It is our sense of time, our sense of ritual, which in the long run creates our sense of place.[24]

This eloquent description evokes the early farmers of eastern Virginia described by Fischer, who shared conceptions of place through the ritualistic practice of court and market days, and the ceremonies of birth, marriage, and death.

The writer and researcher David Seamon calls these ritual repetitions the "place-ballet," which he describes as the everyday activities that produce feelings of existential belonging within the rhythm of life in place, along with the random moments of variety that gives the everyday its sense of meaning and dimension. "Place, in other words, requires both regularity and variety, order and change. Place-ballet is one means by which a place comes to hold these qualities."[25]

Evocative portrayals like these of place's social dimensions reflect how social capital concepts such as bonding capital are embedded within sense of place theories. It is the regular interactions of familiar people in familiar place, and the trust those interactions engender, coupled with a healthy sprinkle of variety, that gives place its social meaning, they suggest. However, not all place constructs are social; an emerging body of research examines the influence of the natural environment in sense of place. Some of this research indicates that people who are more cognitively and emotionally attached to place exhibit more significant concerns about environmental stewardship and protection. These attachments to physical attributes of place may develop faster than social ones because it takes longer to create social relationships than to interact meaningfully with natural settings. Other findings indicate that length of residence and community involvement in an area strongly influence attitudes toward protecting that area's environmental resources. A review of place research published in 2017 indicated that sense of place connections are primarily based on human interactions with the biophysical environment, not social interactions, and that strong people-place relations can strengthen protective norms for the local environment.[26]

The connection between sense of place and environmental stewardship provides intriguing possibilities for policymakers, indicating that place-based policy can lead to more sustainable environmental and land-use practices. However, the positive relationship between place attachments and environmental stewardship may be stronger in what research C.K. Trentleman calls "high amenity communities," which boast pristine or scenic natural environments that tend to attract tourists or affluent homeowners. In "mixed amenity" places where this aesthetic appeal is less strong, or where the landscape has been marred by industrial pollution or environmental disaster, the relationship is often more ambivalent and less conducive toward environmental protections. Indeed, as people's attachment to place becomes less stable and more transitory through economic decline or environmental degradation, sense of place becomes more fragmented and less communal, leading to subsequent declines in environmental stewardship.[27]

The term "agency" refers to the capacity of individuals or groups to make their own free choices, control destiny, or engage in collective action. In the words of researchers Chris Ling and Ann Dale, "social capital is a potential, agency activates it." Agency, then, connotes a more intentional causality toward meaningful change, and both social capital and agency are necessary to affect it—one requires the other, and both are to some degree necessary. Not surprisingly, communities where people feel little connection or control over environmental stewardship feel less agency, resulting in social fragmentation. While social capital is a critical component of agency, it does not always create agency. Therefore, communities or individuals without "agency" are incapable of driving change, no matter the presence of social capital, making agency a critical component of a group's ability to respond to social and/or environmental crisis.[28]

In her work on resiliency, British researcher and author Katrina Brown looks at the role of agency in stakeholder process, positing that transformative response to environmental challenges requires moving from individual, everyday forms of agency to collective and strategic forms.

> If we re-vision resilience and take an agency-oriented perspective, then it provides an opportunity to reconsider the negotiations about desired futures and actions ... This re-visioning puts emphasis on the process of change, on how decisions are made and actions are taken.[29]

Identifying the relationship between agency and capacity for change is perhaps no great revelation. Nor is linking agency to social capital and sense of place. The three are mutually interdependent. For most of us, it makes sense that communities with higher degrees of social capital have more capacity for collaborative change, and that this same power dynamic applies to communities where social and environmental dynamics create a strong sense of place. The dilemma then for policymakers and farmland preservationists becomes activating sense of place and social capital into collective agency given not only the structural barriers facing many "mixed" or "low" amenity communities, or places where the fabric of community has been disrupted by exponential growth and development, but also the general decline in civic participation lamented by Putnam.

The problematic relationship between agency and civic participation in modern society highlights the challenges for practical applications of social capital and place-based theories in "wicked" problems like farmland preservation. We can probably all agree that local farmers need to be active in local government to preserve the rural character of their farming communities. But the "dark side" of bonding capital obstructs this recognition, as many rural communities are still dominated by insular, suspicious mindsets that pose resistance to change and new ideas. And these structural, cognitive roadblocks are often exacerbated for the community's minority or economically disadvantaged members. If there can be no agency without power, people of color and the poor are still regularly excluded from local power structures in American communities. Unfortunately, these power dynamics are often at play in farming communities, where an inherent conservatism and suspicion of outside perspectives create resistance to civic participation.[30]

Cosmic Capital and the Farmland Preservation Ethic

Of course, social capital, sense of place, civic virtue, and agency never come up in the discussions at our agricultural advisory board meetings, even if the social scientist in me is aware of those concepts swirling through the subtext of our discussions. And in the day-to-day work of farmland preservation, I also stay tuned to these ideas, frequently finding that social capital and sense of place are foundational components of the farmland preservation ethic, even if the fragmentation of agency in rural places hinders its application.

In fact, I consider these two components both interdependent and equally necessary for meaningful farmland preservation to occur. My name for this dualistic influence is "cosmic capital," which I define as the human affiliation with a physical place expressed through dimensions of place and community bonds, and the capacity for this affiliation to activate environmental stewardship at both the individual and community level.

The theoretical inspiration for "cosmic capital" is based in Bourdieu's characterization of capital as the capacity of individuals and groups to impact situations in various aspects of life. However, if Bourdieu depicted social capital as a collective pursuit of social power and group interests, cosmic capital also incorporates human connection to place as explanation for the individual or collective motivation to preserve physical landscapes. This emotional or affectionate affiliation with place forms a more altruistic vision of collective action than Bourdieu's instrumental one.

The term "cosmic capital" is inspired broadly by the influence of the early ecologist Alexander von Humboldt, who borrowed the term "cosmos" from early Greek philosophers in his nineteenth-century treatise on the relationship between human and nature titled *Kosmos*. For Humboldt, the cosmos was organized into two aspects, "order" and "adornment." Order refers to the laws of the physical universe that can be observed as patterns and regularities. Adornment, in contrast, is the natural world defined only through human interpretation—a constructivist interpretation of the natural landscape, if you will, the way we also interpret society, culture, and place through our own experiences. To Humboldt, cosmos is both ordered and beautiful, although its beauty is accessible only through the human mind.

Consider sense of place then as the adornment that gives cognitive meaning to our cognitive connections to a physical place such as a farm. We process this adornment through our appreciation of a place's aesthetic appeal, the way the grasses of the field shimmer in the autumn sunset, or the spectral beauty of a tree canopy shading a country stream, the bucolic order of a tidy farmstead with chickens and guinea fowl pecking in the yard.

The "ballet" of our regular interactions with place over time also creates adornment. True, the daily interactions with field, forest, and farmstead are part of this sense, but also a part of it are the daily interactions with family and neighbors that create the social breadth of place. As are the regular interactions of our parents and ancestors, transmitted through generations by narrative and family lore.

And if it is this connection to place that inspires the desire to preserve it in the face of disruptive change, this preservation ethic cannot be activated by sense of place alone. The American "spirit of the law" that demands land be used for economic purpose is simply too strong to counter with place affiliations. Only when sense of place is partnered with social capital can meaningful farmland preservation take place on both the micro and macro scale. The "cosmic capital" paradigm, therefore, posits that sense of place and social capital, in both positive and negative forms—the inward facing ties of bonding social capital based on shared experience and shared values, along with connection to outside resources and expertise created through bridging capital—are instrumental in creating effective farmland preservation policy. Therefore, both are connected through a symbiotic relationship.

In this theoretical model for cosmic capital, a key to its activation is understanding its strength and influence within a specific social grouping, and then using this understanding to promote local farmland preservation policy. Local engagement leveraging local place affiliations to craft farmland preservation policy reflecting local values is the recipe for success, with cosmic capital its special, and most important, ingredient. However, this understanding is often the easiest part of the politics of place. The more challenging part is overcoming the resistance that so often arises in opposition to this type of policy, including the residential development lobby, which fears communal activism and spends big money to suppress it, and "property rights" advocates, who apply Kemmis's "first language" of individualism, using the rhetoric of universal rights to oppose change and adaptation. So, next, we look at governance frameworks that attempt to achieve this activation—the politics of place.

Notes

1 North Carolina General Statutes Article 61, Chapter 106, accessed March 8, 2025, https://www.ncleg.net/enactedlegislation/statutes/html/byarticle/chapter_106/article_61.html.
2 Troy D. Glover and John L. Hemingway, "Locating Leisure in Social Capital Literature," *Journal of Leisure Research* 37, no. 4 (2005): 387–401. Also, James S. Coleman, "Social Capital in the Creation of Human Capital," *American Journal of Sociology* 94 (1988): 95–120.
3 Pierre Bourdieu and L.J.D. Wacquant, *An Invitation to Reflexive Sociology* (Polity Press, 1992), 119.
4 Pierre Bourdieu, "The Forms of Capital," in *Handbook of Theory and Research for the Sociology of Education*, ed. J. Richardson (Greenwood, 1986), 241–258.
5 Robert Putnam, *Bowling Alone: The Collapse and Revival of American Community* (Simon & Schuster, 2000), 113.
6 David Hackett Fisher, *Albion's Seed: Four British Folkways in America* (Oxford University Press, 1989), 392–394.
7 Daniel Kemmis, *Community and the Politics of Place* (University of Oklahoma Press, 1990), 70–71.
8 Alexis de Tocqueville, *Democracy in America*, ed. Harvey C. Mansfield and Delba Winthrop (Chicago University Press, 2000), 182.
9 Tocqueville, *Democracy in America*, 489–490.
10 Tocqueville, *Democracy in America*, 491.
11 Putnam, *Bowling Alone*, 21.
12 Harvey C. Mansfield and Delba Winthrop, "Editor's Introduction," *Democracy in America*, lxvii–lxviii.

13 Rima Dael, "Civic Engagement, Social Capital, and Robert Putnam's 'Bowling Alone': A Critique," National Federation of Community Broadcasters, accessed September 30, 2024, https://nfcb.org/civic-engagement-social-capital-and-robert-putnams-bowling-alone-a-critique-by-a-rima-dael-nfcb-ceo/#:~:text=However%2C%20his%20narrative%20often%20overlooks,how%20gender%20impacts%20civic%20engagement.
14 Putnam, *Bowling Alone*, 23.
15 Carl Folke et al. "Adaptive Government of Social-Ecological Systems," *Annual Review of Environmental Resources* 30 (2005): 441–473.
16 Ö. Bodin and B.I. Crona, "The Role of Social Networks in Natural Resource Governance: What Relational Patterns Make a Difference?" *Global Environmental Change* 19, no. 3 (2009): 366–374.
17 Folke et al., "Adaptive Government."
18 M. Garguilo and M. Benassi, "Trapped in Your Own Net? Network Cohesion, Structural Holes, and the Adaptation of Social Capital," *Organizational Science* 11, no. 2 (2000): 183–196.
19 M. Edwards and J. Onyx, "Social Capital and Sustainability in a Community Under Threat," *Local Environment* 12, no. 1 (2007): 17–30. Also, Maria Lewicka, "Place Attachment: How Far Have We Come in the Last 40 Years?" *Journal of Environmental Psychology* 31 (2011): 207–230.
20 Putnam, *Bowling Alone*, 23.
21 Lewicka, "Place Attachment," 211.
22 Yi-Fu Tuan, *Space and Place: The Perspective of Experience* (University of Minnesota Press, 1977).
23 Tuan, *Space and Place*, 149.
24 J.B. Jackson, "A Sense of Place, a Sense of Time," *Design Quarterly* 164 (1995): 25.
25 David Seamon, "Body-Subject, Time-Space Routines, and Place-Ballets," *The Human Experience of Space and Place* 148 (2013): 163.
26 V.A. Masterson et al., "The Contribution of Sense of Place to Social-ecological Systems Research: A Review and Research Agenda," *Ecology and Society* 22, no. 1 (2017).
27 C.K. Trentleman, "Place Dynamics In a Mixed Amenity Place: Great Salt Lake, Utah," *Human Ecology Review* (2011): 126–138.
28 Chris Ling and Ann Dale, "Agency and Social Capital: Characteristics and Dynamics," *Community Development Journal* 49, no. 1 (2014): 4–20, quote on p. 8.
29 Katrina Brown, *Resilience, Development, and Global Change* (Routledge, 2016), 121.
30 Andrew Waters and Charles Chancellor, "Bridging Capital in Community Conservation: A Case Study of the Taos Community Conservation Plan," *Society and Natural Resources* 35, no. 7 (2022): 745–762.

Bibliography

Bodin, Ö., and B.I. Crona. 2009. "The Role of Social Networks in Natural Resource Governance: What Relational Patterns Make a Difference?" *Global Environmental Change* 19, no. 3. doi:10.1016/j.gloenvcha.2009.05.002.

Bourdieu, Pierre. 1986. "The Forms of Capital." In *Handbook of Theory and Research for the Sociology of Education*, edited by J. Richardson. Greenwood.

Bourdieu, Pierre, and L.J.D. Wacquant. 1986. *An Invitation to Reflexive Sociology*. Polity Press.

Brown, Katrina. 2016. *Resilience, Development, and Global Change*. Routledge.

Coleman, James S. 1988. "Social Capital in the Creation of Human Capital," *American Journal of Sociology* 94. doi:10.1086/228943.

de Tocqueville, Alexis. 2000. *Democracy in America*, edited by Harvey C. Mansfield and Delba Winthrop. Chicago University Press.

Edwards, M., and J. Onyx. 2007. "Social Capital and Sustainability in a Community Under Threat," *Local Environment* 12, no. 1. doi:10.1080/13549830601098206.

Fisher, David Hackett. 1989. *Albion's Seed: Four British Folkways in America*. Oxford University Press.

Folke, Carl, 2005. "Adaptive Government of Social-Ecological Systems," *Annual Review of Environmental Resources* 30. doi:10.1146/annurev.energy.30.050504.144511.

Garguilo, M., and M. Benassi. 2000. "Trapped in Your Own Net? Network Cohesion, Structural Holes, and the Adaptation of Social Capital," *Organizational Science* 11, no. 2. doi:10.1287/orsc.11.2.183.12514.

Glover, Troy D., and John L. Hemingway. 1988. "Locating Leisure in Social Capital Literature," *Journal of Leisure Research* 37, no. 4 (2005): 387–401.

Jackson, J.B. 1995. "A Sense of Place, a Sense of Time," *Design Quarterly* 164. doi:10.2307/4091350.

Kemmis, Daniel. 1990. *Community and the Politics of Place*. University of Oklahoma Press.

Lewicka, Maria. 2011. "Place Attachment: How Far Have We Come in the Last 40 Years?" *Journal of Environmental Psychology* 31. doi:10.1016/j.jenvp.2010.10.001.

Ling, Chris, and Ann Dale. 2014. "Agency and Social Capital: Characteristics and Dynamics," *Community Development Journal* 49, no. 1. doi:10.1093/cdj/bss069.

Masterson, V.A., 2017. "The Contribution of Sense of Place to Social-Ecological Systems Research: A Review and Research Agenda," *Ecology and Society* 22, no. 1. doi:10.5751/es-08872-220149.

Putnam, Robert. 2000. *Bowling Alone: The Collapse and Revival of American Community*. Simon & Schuster.

Seamon, David. 2015. "Body-Subject, Time-Space Routines, and Place-Ballets." In *The Human Experience of Space and Place*, edited by Anne Buttimer and David Seamon. Routledge.

Trentleman, C.K. 2011. "Place Dynamics in a Mixed Amenity Place: Great Salt Lake, Utah," *Human Ecology Review* 18, no. 2: 126–138.

Tuan, Yi-Fu. 1977. *Space and Place: The Perspective of Experience*. University of Minnesota Press.

Waters, Andrew, and Charles Chancellor. 2022. "Bridging Capital in Community Conservation: A Case Study of the Taos Community Conservation Plan," *Society and Natural Resources* 35, no. 7. doi:10.1080/08941920.2022.2078525.

12 The Politics of Place

In *Community and The Politics of Place*, Daniel Kemmis writes about the role of civic participation in American governance and society. Although more optimistic than Putnam in *Bowling Alone*, he urges his readers to embrace the "politics of inhabitation," particularly those in the American West (Kemmis served as the speaker of the Montana House of Representatives and mayor of Missoula, Montana), framing his pleas through a sense of what has been lost since the early twentieth century:

> Citizen alienation, wilderness and farm policy, economic development strategies: all of these topics and dozens of others call into question the way that we in this nation and this region go about the business of being public. Voter apathy is a matter of a deepening denial of our capacity for, or our interest in, being public at all.

In calling out the special role of agricultural policy in this "business of being public," Kemmis evokes Thomas Jefferson's vision of American agrarianism:

> Farm policy raises the issue of how we publicly chose to rank our values ... As we consider issues like open land and agriculture against this background, we naturally recall Jefferson's insistence that an effective self-governing "public" depended upon both agricultural and open land.[1]

Jefferson, Kemmis observes, viewed the role of the farmer not only from a symbolic perspective but also a political one. Jefferson believed the way early Americans made their living influenced the way they participated in government, and the mutual sense of responsibility cultivated in our early farming economies was an indelible feature of America's early republican government. Farmers, Jefferson argued, were better at civic participation because the farm economy relied on the social-ecological interdependence of the agricultural system. But in the commerce and manufacturing that was already beginning to take over America's urbanizing areas, Jefferson feared

DOI: 10.4324/9781003644545-15

> the disconnectedness and anonymity which seemed necessary to accompany them. Jefferson saw clearly that those who made their living through these activities were wholly dependent upon the choices of utter strangers, known only as "consumers." This, he argued time and again, led to the undermining of morals, and particularly civic virtues.[2]

Jefferson feared the vision of Federalism outlined in the United States Constitution, where the primary political entity was the nation itself. This emphasis on a federal government in turn emphasized a commercially competitive, dehumanized civil society, feared Jefferson, who believed a common good could only be achieved in smaller government scales where people engaged in the type of "place-ballet" that engenders a sense of fellowship and mutual responsibility.

In contrast, American Federalists such as Alexander Hamilton were fearful of the instabilities and inherent prejudices in republican forms of government. Instead, they focused on institutionalizing the balancing of interests, creating optimal conditions for an expanding commercial and industrial economy, but abandoning the spirit of civic virtue. "The Federalist plan of government was exactly analogous to Adam Smith's invisible hand, which wrought the highest good in the market even though none of the actors were seeking anything beyond their own individual interest."[3]

This was exactly the type of government Jefferson feared, the type where the individualized rights of the majority outweighed local interests. The political scientist Michael J. Sandel calls this condition the "procedural republic," a public life dominated by a rights-based ethic placing the rights of its citizens above any common conception of the good. This ethic is based in the classic conception of liberalism, Sandel argues, emphasizing individual autonomy, economic and political freedom, and limited government. "The procedural republic ... transforms the relation between liberty and democracy, transforms the relation of the individual and nation-state, and tends to undercut the kind of community on which it nonetheless depends," Sandel writes.

> Liberty in the procedural republic is defined not as a function of democracy, but in opposition to democracy, as an individual's guarantee against what the majority might will. I am free insofar as I am the bearer of rights, where rights are trumps.[4]

Jefferson foresaw these dangers and understood how it threatened the communitarianism he and his fellow Virginia planters enjoyed in the Virginia House of Burgess. The solution to Jefferson's dilemma, proposed by Jefferson's close associate James Madison, was the American geography itself, whose limitless horizons, from the perspective of the eighteenth century, promised a hedge against the tyranny of the majority that Jefferson feared. "Extend the sphere, and you take in a greater variety of parties and interests;

you make it less probable that a majority of the whole will have a common motive to invade the rights of other citizens," Madison wrote in *Federalist Paper Number Ten.* [5]

Jefferson's vision for a nation of "physiocrats" and "yeoman husbandmen" has now passed, leaving us a nation based on the type of commerce, consumerism, and rights-based liberalism he feared. For better or for worse, America has proven itself too vast for Jefferson's republican vision. Sandel posits that the procedural republic began to take hold at the beginning of the twentieth century, as national markets and corporations replaced local economies, leaving the

> decentralized political forms of the early republic outmoded as well. If democracy was to survive, the concentration of economic power would have to be met by a similar concentration of political power. ... We would become more of a democracy only as we became more of a nation.[6]

Yet this concentration has shifted American governance "from a public philosophy of common purpose to one of fair procedures, from a politics of good to a politics of right, from the national republic to the procedural republic."[7]

The challenge for Kemmis, Putnam, and others like them who call for a return to civic participation is to understand how "the business of citizenship" can be conducted in this new American landscape. For Kemmis, this understanding relies on the concept of place. "What holds people together long enough to discover their power as citizens is their common inhabiting of a single place," he writes.[8] The power of this common attachment to place is that its smaller dimensions bring people into relationship with one another, a necessary condition for the cultivation of trust, which itself is an essential component of bonding social capital.

This proximity of place also breeds cooperation, another element of bonding social capital.

> Thus, in a number of settings, people who find themselves held together (perhaps against their will) in a shared place discover as well that their best possibility for realizing the potential of the place is to learn to work together.[9]

This brings Kemmis to his conception of the "politics of place," which he describes as a politics of citizens working out the problems of their place in dialog with one another.[10]

> Such a politics would be a change from what we now know, in many ways. It will depend less upon procedures and bureaucracies and more upon human virtues and patterns of relations. It will be a much less centralized politics ... Cooperation is central to the politics of place, and it will

have to extend to cooperation between right and left, between Democrats and Republicans, even between environmentalists and corporations.[11]

Montesquieu, Tocqueville, and the Politics of Place

Kemmis's ideas on place reflect the political philosophies of Charles Louis de Secondat, Baron de Montesquieu, the widely influential French sociologist of the eighteenth century known to the world as "Montesquieu." Trained as a lawyer, Montesquieu wrote on a wide variety of topics, including governance, law, slavery, and even the influence of climate on human behavior. In both America and Great Britain, however, he is perhaps best known for his prodigious, wildly influential work titled *The Spirt of Laws* (*De l'esprit des lois* in French).

The Spirit of Laws places Montesquieu among the earliest place theorists. Influenced by Aristotle and Plato, Montesquieu's beliefs about the role of geography and climate deeply influenced his theories about place (i.e., the unique characteristics of politics and governance in a particular place, typically a province, city, etc., but also, by extension, a nation), for he believed that a place's economics, morals, and politics must also fit its geography, climate, and people.[12] Contrasting Montesquieu to other Enlightenment philosophers such as Jeremy Bentham and John Locke, the Montesquieu scholar Joshua Bandoch calls his subject a "particularist," who believed that each political order must fit its specific place. "Locke's standards do not change over time and place—they are universal. Bentham's principle of utility is grounded in a universal imperative: to maximize the greatest good for the greatest number of people ... Montesquieu saw that we should study the circumstances of a place—its religion, commerce, laws, institutions, physical environment, and *moeurs* (mores)—to determine the best political order for that place."[13]

For Montesquieu, *moeurs* are a society's collective passions and opinions, as well as its pleasures and tastes. A people's *moeurs* are linked to its feelings, beliefs, and way of thinking. In Montesquieu's writing, *moeurs* are closely related to manners, with *moeurs* describing a society's inner beliefs and values, and manners referring to how the *moeurs* are expressed in outer behaviors.[14]

Like *moeurs*, Montesquieu's concept of *esprit* is also necessary for understanding his writings on place. Broader than his concept of *moeurs, esprit* refers to the common character of the place's people, or its ethos.[15] Montesquieu further differentiates between the general spirit (*esprit general*) and the spirit of the law (*esprit des lois*). The spirit of the law refers specifically to how the *esprit* impacts the place's laws and the product of that impact on its society. Although Montesquieu defined the terms *esprit* and *esprit general* differently throughout his writing, in *The Spirit of the Laws*, he provides the following general definition of the *esprit general*: "Many things govern men: climate, religion, laws the maxims of government, examples of past things, *moeurs*, manners; from this, it results that a general spirit (*esprit general*) is formed."[16] Therefore, the *moeurs* are a component of the *esprit general*, which is extremely complex, formed through the "infinite causes" that shape it over time. The sociologist Raymond Aron describes this process as "the way of living, behaving, thinking, and feeling of a particular collectivity, as geography and history have produced it."[17]

Like Locke and his disciples, Montesquieu believed that government should protect three basic goals: liberty, security, and prosperity. But if Locke believed these goals are universal, and trump other societal objectives, such as community, Montesquieu believed they should be adapted to each particular place. In this way, Montesquieu's politics of place rejected the notion that there is a "best" or ideal order. In politics and governance, Montesquieu believed, one size does not fit all. Religion, law, institutions, commerce, history, climate, *moeurs*, and manners all influenced a place's *esprit*, and thus should be reflected in its governance. This shifts Montesquieu's vision of "good" governance from the "rights-based" framework of Locke and his philosophical descendants reflected in the American "Bill of Rights" to one that focuses on solving the political issues of a particular place, based on that place's particular *moeurs* and *esprit*. And like Sandel, Montesquieu believed governments should steer clear of the kind of "rights-based" approaches demanded by the republic, which present human or "natural" rights as universal. Rather, Montesquieu believed particularistic government should focus on protecting the individual rather than protecting the absolute "right."

But how do governance and policy achieve this particularistic approach, if not guided by the lodestar of natural rights? Bandoch describes the three components of Montesquieu's politics of place. First, policymakers must understand how the variables of place—religion, climate, landscape, history, commerce, etc.—influence and operate within a particular place. Second, they must determine how these variables interact with one another, recognizing that each does not operate in a vacuum separate from the others. In the final step, policymakers must understand how the variables interact in a particular place to form the *esprit general* and the *esprit des lois*. [18]

Montesquieu's perspective on the politics of place obviously creates issues for modern governance. He pays scant attention to issues of equity and justice, and the contrarian could argue that if the *esprit* of a particular place was discriminatory or inherently racist, then the governance of that place should accommodate its racism, at least according to a strict interpretation of Montesquieu's theories. But Montesquieu believed in the absolutes of liberty, prosperity, and security, and a government that fails to support the pursuit of these absolutes for its minority members presumably fails in these requisites. His writing also condemned despotism and slavery, which he viewed as contrary to the essence of human nature and the universal value of liberty.[19]

But Montesquieu was writing in the eighteenth century, and his positions on equity and human rights are not as unconditional as the twenty-first-century ethicist might prefer. This should not distract us from the *esprit*, if you will, of Montesquieu's view: all politics are particular and should be suited to the place's *esprit,* within some basic overall principles. And if this view is at odds with today's "procedural republic," along with progressive views on many issues that define American politics today—such as abortion, transgender rights, and the right to bear arms—it is still a useful principle for the idea of farmland preservation, which thankfully avoids (for the most part) such contentious political divisions.

But what about community? If Montesquieu believed in the pillars of liberty, security, and prosperity as political absolutes, where does community fit under that foundation, and what is government's responsibility toward fostering it? Bandoch argues that Montesquieu believed a desire for communal participation is what drives humans toward society. "Montesquieu does not affirm the existence of rights, as Hobbes and Locke do. Montesquieu does not say that man enters society to protect these rights. Rather, Montesquieu argues that men naturally desire to enter society."[20] Once in society, humans naturally expect the government to protect the rights upon which society functions, but it is society itself, the communal interaction of commerce and fellowship (i.e., the "place-ballet"), that is our primary motivation for accepting its laws and government. Community, therefore, is the reason for government, not a byproduct of it.

Montesquieu died in 1755 having never visited the America colonies, and so gave us little insight into how his theories apply to American culture. However, his disciple Alexis de Tocqueville wrote extensively on American culture and society in his masterwork, *Democracy in America,* after he visited the country from 1831 to 1832, and so gives us a perspective on the United States through the prism of Montesquieu's ideology.

Tocqueville did not directly address the role of community in *Democracy in America,* but he did observe the important role of localism and voluntary civic activity in America's *esprit general.* Tocqueville called these characteristics America's "equality of condition," but he did not use the term in a modern sense, suggesting that all people deserve the same rights. Rather, he was referring to the horizontal character of the American class system, unencumbered (at least to some degree, and perhaps more so in the frontier towns of the Mississippi basin where he visited and formed his theories than its eastern metropolises) by the hierarchies and prejudices of European caste systems.

"Among the novel objects that attracted my attention … nothing struck me more forcibly than the general equality of condition among the people," Tocqueville writes in an early passage in *Democracy in America.* Tocqueville's conviction that Americans were immune to Europe's class prejudices was central to his understanding of its civil society—indeed he believed it was one of his defining characteristics of American society, even if it ignored the class distinctions that existed both then and now.

> I readily discovered that the prodigious influence of this fact extends far beyond the laws of the country, and that it has no less effect on civil society than on the government; it creates opinions, gives birth to new

> sentiments, and modifies what it does not produce. The more I advanced in the study of American society, the more I perceived that this equality of condition is the fundamental fact from which all others seemed to be derived.[21]

Yet Tocqueville understood and feared the influence of a "procedural republic" on communal aspects of American life. "When social conditions are equal, every man is apt to live apart, centered in himself and forgetful of the public," he wrote,[22] even if his travels assured him that American localism could mitigate the isolating impulses of personal freedom and the pursuit of self-interest into the future.[23]

As Tocqueville traveled through the small towns of America, he believed the civic activities he saw there—in particular, the New England town meetings—provided a moderating influence on American individualism.

> No sooner do you set foot upon American ground than you are stunned by a kind of tumult, a confused clamor is heard on every side, and a thousand simultaneous voices demand the satisfaction of their social wants. Everything is in motion around you; here the people of one quarter of a town are met to decide upon the building of a church; there the election of a representative is going on; a little farther, the delegates of a district are hastening to the ton in order to consult upon some local improvements; in another place, the laborers of a village quit their plows to deliberate upon the project of a road or a public school.[24]

In other words, Tocqueville believed a defining characteristic of American culture was precisely the type of civic participation whose loss Putnam laments in *Bowling Alone.* But Putnam suggests this loss can be regained by a return to Tocqueville's model of self-interest served by collective activity. "As Tocqueville pointed out, however, American democracy worked not because Americans obeyed some impossibly idealistic rule of selflessness, but rather because we pursued 'self-interest rightly understood,'" Putnam writes. "Members of a community that follows the principle of generalized reciprocity … find that their self-interest is served."[25]

Or, as the Tocqueville scholar John Ehrenburg puts it,

> Tocqueville's appreciation of localism drove his analysis of the American tendency to voluntarily organize. Indeed, his signature contribution to contemporary theories of civil society is conditioned by his respect for the way American small-scale, parochial activities fuse the private strivings of self-interested individuals with the necessitates of cooperation and community. Americans had learned how to reconcile equality of conditions and political democracy with freedom because their powerful traditions of local autonomy kept the modern state at bay.[26]

From this broad palette of self-interest, community, and localism, let's now return to Kemmis, who consolidates these themes in the closing chapters of *Community and the Politics of Place*.

Like Montesquieu, Kemmis believed place is the essential component of politics, but that place is based on the relationships of social interactions (and we can presume, the social capital those relationships produce), which can only occur within that place's confines. Speaking collectively in a community sense, Kemmis writes, "What 'we' do depends … upon how we choose to relate to each other, to the place we inhabit, and the issues which that inhabiting raises for us."[27]

Kemmis does not ignore the influence of geography on the *general esprit* of place, or its influence on other social and symbolic elements of community. Already we have read how eastern Montana's harsh landscape influenced his perspectives. "Rugged country, then, has always selected more than its share of rugged individualists … the harder the country, the more fiercely it selects its inhabitants." But if this rugged individualism was a defining *esprit* of Kemmis's childhood, the interdependence of neighbors was another, and that "everyone was needed by everyone else in one capacity or another."[28]

These lessons endured as he entered a career of politics and small-town governance, and at the center of Kemmis's vision for the politics of place is the concept of social capital—the idea that people, working together collaboratively, create meaningful place-based policy, not just compromises. In the final chapter of *Community and the Politics of Place,* Kemmis spends time developing this idea of a "practice of citizenship." In such a practice, citizenship moves beyond mere acquiescence to a place's rules and laws, or perhaps occasional attendance at a public meeting on an issue or initiative of interest, to one of engaging regularly and consistently in political process.

> Rules and laws certainly have their role … but in the end they are, by their very nature, abstractions from the particular … practice arrive at the common through the particular, through a shared way of arranging and relating to the actual, physical—"our world."[29]

This "practice of citizenship" is based on the premise that political decisions regarding community issues such as farmland preservation must take place in a dialog of citizens encompassing a diverse range of opinions and viewpoints. Only through the friction of this process can the trust necessary to implement solutions to "wicked problems" like farmland conversion develop. His argument is essentially one of stakeholder process: political decisions regarding community issues such as farmland preservation must take place in a dialog of citizens encompassing a diverse range of opinions and viewpoints.

A central component of this practice is the "second language of commitment"—the way we speak and treat one another that acknowledges the shared responsibilities of place, and the relationship between people and of people with land and place. And only through the scale of local place is this type of relationship-building possible through the practice of building social capital. In fact, it is only in the emotional and physical attachment to a common place that this type of collaborative politics can flourish. "Places breed cooperation and out of this ancient relationship of place to human willing, that specific activity which is rightly called the 'politics' is born."[30]

Kemmis has specific ideas about the scale at which the politics of place can occur. This idea of scale plays an important role in his theories on rural life and incorporates economic dimensions into his vision for the politics of place. For Kemmis, the "city-state" or "city-region"—the city and its surrounding rural area, or what he calls the city's "hinterlands"—is the ideal political scale for the politics of place. He argues that standard units of American government such as the nation, state, and county too often "cut across natural units of inhabitation, leaving inhabitants cut off from each other in terms of their capacity to act together politically—to will a common world."[31]

The city-state is a more natural unit of governance because it entails the symbiotic relationship between rural and urban created by both a common economy and political order. Central to his theory on political organization is the concept of import substitution, an idea he borrows from the political theorist Jane Jacobs in her book *Cities and the Wealth of Nations*. There she argues that replacing imported goods with locally produced goods is what makes local economies innovate, grow, and thrive. "Economic life develops by grace of innovating; it expands by grace of import-replacing. These two master economic processes are closely related, both being functions of a city economy," Jacobs writes there. In contrast is what Jacobs calls the "national economy," based on the premise that, in the age of technological advancement, industrialization, and globalization, national economies have become the defining units of economic life. But national economies lead to passive local economies that do not create economic change themselves but instead respond to the national (and increasingly international) one.[32]

Kemmis's application of Jacobs' economic theories to the city-state—the city and its hinterlands—expands the concept of the city-state from a natural economic ecosystem to a natural social ecosystem as well.

> If cities, in relation to their hinterlands, have the capacity to define working economies, then it makes sense that the same city regions with constitute economies should also be "city-states," or the ecosystem within which 'the task of willing a common world makes sense."[33]

In the chapters that follow, we look at some new directions in farmland preservation that focus on farmland preservation through the lens and scale of the politics of place. Not all of these policies and programs are local in nature. Some are operated through a statewide or regional administration. Some harness the power of the Internet to cultivate financial investors from around the country. But at their core, all are focused on farmland preservation not from a resource agrarian perspective—the idea that saving farmland is somehow an issue of national security or national sustainability—or approaches that focus through broad planning or zoning approaches, applying the same standards of restrictions to all properties in a region, no matter their individual use or characteristics. Instead, farmland preservation in the politics of place focuses on a perspective of supporting local farm communities through cultivating local farmers—what some on the farmland preservation movement call "agricultural viability"—and giving these farmers opportunities to keep the American farming tradition alive within their local communities. It is a politics of place for a new generation, using new tools within the American *esprit general*.

Notes

1 Daniel Kemmis, *Community and the Politics of Place* (University of Oklahoma Press, 1990), 41–42.
2 Kemmis, *Community and the Politics of Place*, 20–21.
3 Kemmis, *Community and the Politics of Place*, 15.
4 Michael J. Sandel, "The Political Theory of the Procedural Republic," *Revue de Métaphysique et de Morale* 93, no. 1 (1988): 67.
5 James Madison, *The Federalist*, Paper Number 10 (Liberty Fund, 2001), 48.
6 Michael J. Sandel, "The Procedural Republic and the Unencumbered Self," *Political Theory* 12, no. 1 (1984): 92.
7 Sandel, "Procedural Republic," 93.
8 Kemmis, *Community and the Politics of Place*, 117.
9 Kemmis, *Community and the Politics of Place*, 122.
10 Kemmis, *Community and the Politics of Place*, 123.
11 Kemmis, *Community and the Politics of Place*, 138–139.
12 Jessica Barnes and Michael R. Dove, "Introduction," in Jessica Barnes and Michael Dove, Eds., *Climate Cultures: Anthropological Perspectives on Climate Change* (Yale University Press, 2015), 30.
13 Joshua Bandoch, *The Politics of Place: Montesquieu, Particularism, and the Pursuit of Liberty* (University of Rochester Press, 2017), 1–2.
14 Céline Spector, "Customs, Moeurs, Manners," A Montesquieu Dictionary, accessed October 18, 2024, https://dictionnaire-montesquieu.ens-lyon.fr/en/article/1376474087/en/#:~:text=The%20m%C5%93urs%20are%20linked%20to,sanction%20(XIX%2C%2014).
15 Bandoch, *Politics of Place*, 28.
16 Excerpted from Bandoch, *Politics of Place*, 27; attributed to Montesquieu, *The Spirit of the Laws*, XIX, 5, 558.
17 Raymond Aron, *Main Currents in Sociological Thought*, Vol. I (Transaction Publishers, 2008), 47.
18 Bandoch, *Politics of Place*, 24.
19 Aron, *Main Currents*, 39–40.
20 Bandoch, *Politics of Place*, 20.
21 Alexis de Tocqueville, *Democracy in America*, ed. Harvey C. Mansfield and Delba Winthrop (Chicago University Press, 2000), 3.

22 Tocqueville, *Democracy in America*, 256.
23 John Ehrenberg, *Equality, Democracy and Community from Tocqueville to Putnam* (New York University Press, 2002), 54.
24 Tocqueville, *Democracy in America*, 249–250.
25 Robert Putnam, *Bowling Alone: The Collapse and Revival of American Community* (Simon & Schuster, 2000), 135.
26 Ehrenberg, *Equality, Democracy and* Community, 54.
27 Kemmis, *Community and the Politics of Place*, 41.
28 Kemmis, *Community and the Politics of Place*, 44, 71.
29 Kemmis, *Community and the Politics of Place*, 122.
30 Kemmis, *Community and the Politics of Place*, 122–123.
31 Kemmis, *Community and the Politics of Place*, 120.
32 Jane Jacobs, *Cities and the Wealth of Nations* (Penguin, 1984), 39, 34.
33 Kemmis, *Community and the Politics of Place*, 121.

Bibliography

Aron, Raymond. 2008. *Main Currents in Sociological Thought*, Vol. I. Transaction Publishers.

Bandoch, Joshua. 2017. *The Politics of Place: Montesquieu, Particularism, and the Pursuit of Liberty*. University of Rochester Press.

Barnes, Jessica, and Michael R.Dove. 2015. "Introduction." In *Climate Cultures: Anthropological Perspectives on Climate Change*, edited by Jessica Barnes and Michael Dove. Yale University Press.

de Tocqueville, Alexis. 2000. *Democracy in America*, edited by Harvey C. Mansfield and Delba Winthrop. Chicago University Press.

Ehrenberg, John. 2002. *Equality, Democracy and Community from Tocqueville to Putnam*. New York University Press.

Jacobs, Jane. 1984. *Cities and the Wealth of Nations*. Penguin.

Kemmis, Daniel. 1990. *Community and the Politics of Place*. University of Oklahoma Press.

Madison, James. 2001. *The Federalist: "Paper Number 10."*Liberty Fund.

Putnam, Robert. 2000. *Bowling Alone: The Collapse and Revival of American Community*. Simon & Schuster.

Sandel, Michael J. 1984. "The Procedural Republic and the Unencumbered Self," *Political Theory* 12, no. 1: 92. doi:10.1177/0090591784012001005.

Sandel, Michael J. 1988. "The Political Theory of the Procedural Republic," *Revue de Métaphysique* et de *Morale* 93, no. 1: 67.

Spector, Celine. "Customs, Moeurs, Manners," A Montesquieu Dictionary. Accessed October 18, 2024. https://dictionnaire-montesquieu.ens-lyon.fr/en/article/1376474087/en/#:~:text=The%20m%C5%93urs%20are%20linked%20to,sanction%20(XIX%2C%2014).

13 New Tools for Old Paradigms

If the past fifty years of farmland preservation have taught us a single lesson it is that American farming is not one "community" of agriculture. The needs and issues facing the farmers of Iowa and Nebraska are different from the needs and issues facing farmers in North Carolina and Maryland, just as those needs and issues are different for the farmers in the eastern part of Chatham County, where residential housing development runs rampant, and western Chatham County, where an elderly generation hangs on to the traditions of cattle and hay farming while the economics of small-scale farming turn bleaker.

True, the success of the conservation easement suggests that all farmland preservation can look the same, while the proliferation of state and local Purchase of Agricultural Conservation Easement (PACE) programs—twenty-nine active statewide PACE programs in 2025, according to the most recent American Farmland Trust survey, with at least 108 independently funded local PACE programs operating in twenty-two states[1]—bolsters that illusion. As we have read, there is good reason for this illusion: the conservation easement's compatibility with the American spirit of the law makes it a convenient farmland preservation tool in an increasingly divisive political age.

But not all American farming communities have access to the considerable funds required to establish PACE programs. And the permanent and heavy-handed restrictions of conservation easements make them a poor fit for all but a small fraction of American farmers, typically the oldest generation, owners of family land handed down through generations facing development pressure in urbanizing areas. This makes conservation easements an awkward fit for a new generation of farmers desperate for access to farmland, or landowners who need more flexible preservation options to accommodate complex family and business dynamics or innovative new sustainable farming practices.

Indeed, finding ways to provide access to farmland and viable farming opportunities may be the central challenge for farmland preservation in the 21st century, or at least the key to keeping family farming alive. And many places in America are responding with new farmland preservation programs and policies focused on these access and viability issues. From a "politics of

DOI: 10.4324/9781003644545-16

place" perspective this focus makes sense; even if a place "preserves" all of its farmland, if there is no one to farm, no one to attend its schools or join its churches or shop at its feed store, no one to shop at its farmers markets, then that place still becomes an empty, barren landscape, preserved farmland or not.

It is this new paradigm of farmland preservation based in the politics of place—combining traditional farmland preservation tools with emerging models for fostering new generations of farmers and enhancing markets for local agriculture—that occupies our attention for the last third of this book. In this chapter we look at ways local communities, often supported by local or national conservation organizations, are moving beyond the mixed records of zoning and planning to launch a new generation of farmland preservation programs centered around "Buy-Protect-Sell" models that provide assistance to small-scale farmers priced out of the traditional farmland real estate market. This chapter also explores an effort to revive the cooperative or "commons" model for farming, an idea steeped in the progressive agrarianism of the late nineteenth and early twentieth century that one organization is looking to revive in the face of agriculture's modern challenges.

Buy-Protect-Sell

Within the American farmland preservation movement is a growing realization that farmland preserved through conservation easements rarely translates into opportunities for new and beginning farmers, nor into opportunities to keep farming viable on farmland that is permanently preserved. Gabrielle Roesch-McNally, the director of American Farmland Trust's (AFT) "Women for the Land" initiative, which supports women landowners and farmers facing barriers on agriculture, says:

> The land access pieces is kind of the elephant in the room for a lot of folks in our work, especially for the new and beginning farmers and also for minority farmers, with the issue of heirs property and other structural issues that have impacted people's ability to access land. It's a huge issue, and there is no silver bullet.[2]

For Jennifer Dempsey, at AFT's Farmland Information Center, the access issue speaks directly to the ability to keep agriculture viable in local communities. "We are really thinking about how to support communities in starting to plan for agriculture, and to plan for not just permanent land protection but then also at the same time really creating opportunities for agriculture," she says.[3]

Holly Rippon Butler is a sustainable agriculture consultant who co-wrote a guidebook titled "Finding Farmland" for the National Young Farmers Coalition (NYFC), a nonprofit focused on providing support and advocacy for young and minority farmers. During her tenure with NYFC, Rippon-Butler's work focused on the land access issue.

> The core of it was really the same—that young farmers were just struggling to find land that they could have long-term security on. The way to create that security is through ownership, and ownership is unaffordable in a lot of cases for young farmers.[4]

David Howard, policy director at NYFC, links the land access issue directly to the viability of local and regional food systems, echoing Kemmis's emphasis on the importance of the "city-state" in the politics of place. "The pandemic showed us how important those local and regional food systems are. But the land access issue persists," he cautions.

> Young farmers are struggling to afford secure land tenure. So even if they're able to lease land, it's an uncertain relationship. And you don't just show up on farmland and start making money on day one. You have to invest in the soil, you have to invest in the infrastructure. So how are you supposed to make those investments if you're leasing land, and it could be taken away from you at the end of the lease?[5]

In their narrative review, Liz Carlisle and her research partners argue that sustainable agriculture is among the "most urgently needed work in the United States" due to the country's looming health, environmental, and rural economic crises. But they found that new entrants for sustainable agriculture struggle for a variety of reasons, ranging from limited markets, expensive equipment, water scarcity, farm labor shortages, and lack of technical training. But land access was among the most prominent barriers, their research determined.

> First are the hurdles associated with finding available and affordable acreage … Second are the historical and ongoing social relations that govern a farmer's ability to derive benefits from arable land … Farmland loss, ballooning agricultural land values, and precarious tenant arrangements all contribute to a land access dilemma for new entrant farmers.[6]

"Buy-Protect-Sell" (BPS) is a relatively new model of farmland preservation that uses the traditional tool of the conservation easement to address the farmland access issue. It is part of a wave of farmland preservation programs that realize creating the type of market opportunities that can keep small and medium-sized farming viable is just as important as the permanent farmland protections that have dominated the "resource agrarian" perspective for the last fifty years.

In typical BPS models, a nonprofit or government organization buys farmland on the open market, and then conducts a recruitment or application process for farming tenants, typically prioritizing candidates who are beginning and/or minority farmers who would otherwise struggle to access viable, affordable farmland. Meanwhile, the organization pursues the sale of a working lands conservation easement—usually through local, state, or federal PACE programs—in order to recoup a portion of their initial investment in the farmland. With the "market value" of the farmland now reduced by the conservation easement, the organization then enters into a purchase agreement with new tenants that enables the farmers to acquire the land at that reduced market value while allowing the organization to recover the remainder or majority of its initial investment. True, the new owners' farm operations are limited by the terms of the conservation easement, but these terms are fully disclosed, and the BPS organization attempts to match the restricted land with farmers whose business plans are a good match. Many of these programs also include some training for the farm candidates, often focusing on business management for smaller-scale farms.

Ann Arbor, Michigan, is one local community experimenting with the BPS model. Home of the University of Michigan's primary campus, Ann Arbor is a rapidly urbanizing community with a population of 120,000 according to 2023 U.S. Census Bureau figures. The city's Greenbelt Program was established in 2003 to help preserve Ann Arbor's rural character and funded through a special property tax assessment approved for a thirty-year term by a public referendum in the same year. According to organizational documents, the majority of the Greenbelt Program's work has been accomplished through the purchase of development rights and partnerships establishing publicly accessible recreation lands. As of 2024, that funding support has translated into ninety-three conservation easement acquisitions, including seventy-five working farms within the Greenbelt boundaries. In return for the funding the city has invested in open space preservation through its Greenbelt program, Ann Arbor has secured $40 million in matching grants and contributions.[7]

"From the early 2000s to 2015, we saw the establishment and growth of the effort to press land affordability and access as a social problem that exists within America's farm system," notes Remy Long, Ann Arbor's deputy director for natural area preservation. Although the Greenbelt Program struggled to find effective solutions for affordability and access issues, Long recalls:

> the phone calls kept coming, the emails kept coming, the meetings kept coming, to the point where we could not ignore that there was a dense enough contingent of land seekers in our area … to recognize we have to be more proactive and start from the beginning of a transaction developing towards that final outcome.

To quantify this demand, the Ann Arbor Greenbelt Program partnered with a branch of Michigan State University Extension to survey new and beginning, socially disadvantaged, and historically underserved farmers in their community about their land access and affordability needs, garnering 180 responses. The results indicated that majority of these farmers were struggling to access affordable farmland or felt little security on the land they currently leased. And most indicated they would be amenable to farming and eventually acquiring "protected" farmland.

Armed with this survey data, Greenbelt staff approached Ann Arbor's city council in 2020 with a proposal for a BPS initiative, and the program recently completed its first BPS project. Both Long and his associate Rosie Pahl Donaldson, a land acquisition specialist who manages the Greenbelt Program, give credit to the city council, who permitted staff to modify the Greenbelt's Program guidelines in order to launch the program.

Finding and acquiring the land was a challenge. In order to make the project work financially, Long and Pahl Donaldson needed to find a tract for sale within the Greenbelt's jurisdictional territory (extending five miles from Ann Arbor's city limits) that could be subdivided into two farm parcels. Each of the two subdivisions would need road access and have to meet other county subdivision and ordinance requirements. Each would have to accommodate a homesite for the eventual farmer/owner with associated septic infrastructure but also meet requirements for a future conservation easement.

Eventually they succeeded, acquiring a 54-acre tract that was subdivided into two tracts of 20 and 34 acres. Funding for the acquisition came from Greenbelt revenues (the dedicated property tax millage) and a matching contribution from the Washtenaw County Parks and Recreation Committee, which also has a natural areas preservation program funded by local property taxes. Next came the equally challenging process of developing eligibility criteria and guidelines to select tenants for their two tracts. In order to support the program's success, program partners wanted to prioritize farmer candidates who already had some experience in farming and could prepare a viable farm business plan. Also, they wanted to prioritize candidates who didn't currently own farmland (or who could demonstrate that their current land access or ability to farm was not secure) but also demonstrated financial ability to eventually purchase the property.

Developing an application process that both addressed the applicants needs and emphasized these qualifications was a challenge, Long and Pahl Donaldson report. "There were all these things to consider about being fair and accurate but also creating measurable evaluations that are not too invasive," Long recalls. "You know, there's a reasonable amount of due diligence that we have to do on our side, but there's also just kind of endless due diligence you could do to make sure that someone is following the rules."

Another challenge was setting guidelines for the property and its use that permitted development of a farm business while still maintaining the program's preservation goals. "It's very much different from traditional conservation easements, where everything's pretty much already specified," explains Pahl Donaldson. In order to accept the restrictions and expectations of the conservation agreement, applicants had to sign a "sale acknowledgment" that disclosed the terms and conditions associated with the ownership transfer. Through this acknowledgment, applicants agreed they would convey a conservation easement back to the Greenbelt Program and their partners in Washtenaw County after they acquired the property.

Despite these challenges, the Greenbelt Program received several applications for the two tracts and eventually selected two "winners." One was a local organic farm that had been in operation for five years and was a well-known vendor at local farmers' markets. Acquiring the BPS property allowed these farmers to move their operation off of leased land, where they had limited long-term security. The other was a native plant and seed operation that also had been in business for five years. Greenbelt Program staff helped these farmers obtain financing from the Farm Service Administration. When the farmers closed on their new properties, they conveyed conservation easements on the properties as part of the same transaction.

Long and Pahl Donaldson agree there were many lessons learned in their first BPS project. In terms of acreage conserved, the project provided a relatively small result in return for all the due diligence and planning it took to complete. But Pahl Donaldson reports the effort was worth it.

> While Buy-Protect-Sell is a far more intensive acquisition process for staff to run and will not generate the same number of acres annually protected, those secondary and tertiary benefits of supporting local farmers and our local food system will be more pronounced if we take this innovative approach.

Noting that the original thirty-year term of Ann Arbor's Greenbelt Program is set to expire in 2033, Long believes the program's successful history helped pave the way for them to move into the BPS arena.

> If we had tried this back at the beginning of the Greenbelt Program, I think the voters and honestly some of the farmers in the area might have been less supportive, but now that we have something to stand on, the success and expediency we've shown leveraging our dollars with external funding, it is now both financially and socially responsible for us to use this innovation for the remainder of the years of the program.[8]

The Conservation Fund is a national conservation organization that has a long record of bringing innovative financing and acquisition strategies to broadscale land conservation projects. It is now bringing that experience to BPS approaches in select American communities through its "Farms Fund" program. So far, the program is working in the Atlanta, Chicago, and Charlotte metro areas, with plans to soon expand into northwest Arkansas. Like the Ann Arbor Greenbelt, the Farms Fund targets local farmers selling within the local food system who face barriers to long-term, secure farmland access with BPS opportunities. "Really what these farmers need is not just FSA loans and the other conventional tools that are out there, but also support in that transition to make their farm business be successful," explains Krisztián

Varsa, the director of the Farms Fund. "A lot of these farmers are leasing on small acreage. They don't really have a path to ownership ... but there isn't a patient path that gets them to ownership in a way that's affordable."

The Farms Fund's first project was working with the "Love Is Love Cooperative Farm," a certified organic vegetable, fruit, and flower farming operation originally established in 2008 that is now a cooperative of five partners/owners. After many years of operation on the outskirts of Atlanta, the business had a strong local clientele through its Community-Supported Agriculture (CSA) program, a business model where local customers "subscribe" to a local farm for regular produce and farm product deliveries. However, despite their established success, the business still struggled to secure land access and financing for growing their business. Through the Farms Fund program, the Conservation Fund acquired a 68.7-acre farm tract in Mansfield, Georgia, a community on the outskirts of Atlanta where Varsa says "subdivisions are starting to creep in."

The Love Is Love Cooperative now has stable, secure access to the farmland while the Conservation Fund seeks acquisition funding for a permanent farmland conservation easement on the property. The cooperative was also able to raise $817,000 for farm infrastructure through a crowd-sourcing initiative, funded largely from local investors.[9] Since that first project, the Farms Fund has created access to farmland for twenty-nine farm owners on 1,300 acres across sixteen different farm properties and facilitated the direct investment of $409,000 in on-farm infrastructure as of mid-2025.[10]

Helping farmers find access to venture capital and other forms of financial support is an important part of the Farms Fund's model, explains Varsa. "The way that we describe the work that we do is 'Buy-*Support*-Protect-Sell,'" he explains. "That really comes from a recognition that these first-generation farmers, next-generation farmers, farmers that want to scale their existing production, don't really have the systems in place for successful local food system production."

One of the key elements of the program is picking farmers who are ready for the Conservation Fund's support through experience and training. "The farmers we work with have to have a certain amount of actual farm management experience, or something to show that they can manage or own a business," says Varsa. As a preliminary screening process, the farmer candidates have to have good credit and provide references. They also need a business plan, although the Conservation Fund will provide them access to a business planning consultant if necessary. Once the farmer or farm business passes the initial screening system, the program works with them to find a specific piece of property that suits their business model. "I think that's one of the secret sauces for our model," says Varsa.

> We're not going out there and grabbing any farm. We take them through the application process. They will become approved within our system. And then working with that farm business, we will try to find the right farm to match that business.

Once the farmer is selected and then matched with a piece of property, the "Support" part of the program kicks in. Using their connections in the conservation field, program staff can help farmers access venture capital.

> A lot of these farmers are looking to grow, which means they are also looking to invest, which means they need infrastructure and equipment and tools ... So, we find effective capital partners that can meet these kinds of farmers where they're at in the process of scaling a new business.

An important part of that process is connecting farmers with NRCS and other technical assistance providers like the local soil and water conservation districts. Varsa tells the story of one female farmer who didn't believe she qualified for NRCS assistance.

> Fast forward three years, this business has a $30,000 cost share contract to do high tunnels and water capture. But she didn't even think she belonged in that system, which is what a lot of small, young, new farmers are told and feel. As so as an entity that has experience and resources and knowledge about how that system works, we're able to push that door open wide enough for the farmer to walk through. So that's another piece of the puzzle that we bring to the system.

Compared to the Ann Arbor BPS program, the Farms Fund program takes a more hands-on approach. Both programs carefully screen their BPS applicants, seeking farmers who have practice and understanding for running a farm business but otherwise face barriers to owning land. Unlike the Ann Arbor program, which held an open proposal process for land it already acquired, the Farms Fund identifies farmers/partners first, and then works with them to acquire a piece of land that meets their farming objectives. With the resources and capacity of a national conservation organization, the Farms Fund helps provide access to capital and technical assistance for their farmers/partners and takes an active role in selling a conservation easement to local PACE programs before transitioning ownership to the new landowners.

But both programs share the same objectives—not only helping local farmers access land but supporting the local food system that is an essential element of farmland preservation. Through their work with local farmers, agencies, and funders, they are also bolstering a "politics of place" that brings attention and resources to their local food system and even empowers that system to address local agricultural viability initiatives, opening doors for new entrants into that system while organizing its components around the farmland access issue, all while developing social capital as an important byproduct. And it attempts all this with conservation easements and adaptations of traditional farmland preservation models.

"Local matters to people because it brings the food closer to them ... and a growing number of people want to feel connected to their farmer," explains Varsa.

> We're [the Farms Fund Program] really focused on building these local food systems around people very intentionally ... it's not just about the land, it's also about the people, food access, fresh—fresh, healthy, local food access that goes beyond the transaction component with the farmer.[11]

The Farmers Land Trust: Looking for New Benefits in a Classic Model

Since the beginning of the American experiment, the country's agriculture system has operated on the ideological premise of single ownership. By an extension of Locke's philosophy, property ownership endowed American landowners—primarily its early farmers and planters—with the inalienable rights that would later become a foundation of our revolutionary principles. Indeed, during its colonial era, only property owners had a right to vote in elections, a practice that continued in America until 1856, when North Carolina became the last state to eliminate property requirements for White male voters in state elections.

Thomas Jefferson touted the democratic advantages of property ownership, believing that, through property ownership, the farm owner was better endowed with the virtue and responsibility necessary to implement the American democratic experience than others. Although Jefferson's "ideals" have long-since been exposed for their inherent prejudices, the ideal of land ownership still looms large in the American *esprit*, even among a new generation of farmers long shed of conventional, institutional thinking.

But collective ownership also has a long history in American agriculture, and among some elements of the farmland preservation movement, it is an idea under revival. Inspired by visions of community shared by social capital proponents like Kemmis and Putnam, the farmland commons movement is an emerging (or re-emerging) approach to farmland preservation that treats farmland as a shared resource to be owned, managed, and stewarded for the benefit of both current and future generations.

"Commons" models of farming and agriculture were essential features of land tenure in America's early colonial period. Agriculture at the Jamestown settlement in Virginia, established in 1607, began as a communal enterprise, and agricultural commons were an integral component of most early settlements in New England, not to mention the established common lands of early Spanish settlements in Mexico and the American West, as well as the indigenous commons across North America pre-dating European settlement.

Western European settlers, however, showed an increasingly zealous inclination toward private property ownership, along with the social and political benefits such ownership conferred. Even though "commons" grazing practices remained well into the nineteenth century, American private property regimes eventually replaced commons models, at least in terms of land ownership.[12] The "progressive" farmer movement of the late nineteenth and early twentieth centuries revived cooperative farming models, but primarily in the arenas of equipment ownership and collective marketing, not land ownership. Following the Civil War, the Grange's embrace of "cooperation in all things" led to the formation of hundreds of agricultural marketing and purchasing cooperatives, along with cooperative stores for retail goods.[13]

The modern agricultural commons movement was influenced deeply by the "New Agrarianism" of the latter twentieth century, as agricultural critics such as Aldo Leopold and Wendell Berry advocated for more sustainable and equitable agricultural models to combat the excesses of industrial agriculture. By the early 2000s, some of these New Agrarians were experimenting with adapting the commons idea to the same nonprofit model that had been employed so effectively by the land trust movement in the 1990s, leading to the organization of "community land trusts." These community land trusts are nonprofit organizations that own and manage land for the benefit of a community. Although frequently oriented around affordable housing initiatives, the model can also be applied to agriculture in the "farmland commons" movement. "The farmland commons model utilizes limited-scope land holding nonprofit structures to hold title to real estate," explains Ian McSweeney, a co-founder and co-executive director of the Farmers Land Trust, a national organization dedicated to promoting the farmland commons model. "The nonprofit model holds title to real estate and then leases it out under secure, equitable, long-term lease structures." The model typically employs long-term ground leases, allowing farmers to build and own housing and farm infrastructure within the commons property while leasing the farmland at affordable rates. "It accounts for equity, security, and long-term affordability for the farmer," explains McSweeney. "So it's addressing that number-one barrier to farming that farmers typically identify—access and security on land."

Like conventional land trusts, farmland commons organizations are focused on preserving and stewarding agrarian working lands. But their mission is focused on providing farmland access to their long-term tenants, not protecting the land itself. "Many times, these farmland commons are created in partnership with conservation land trusts," says McSweeney.

> So conservation easements are a part of our work. But our model was created in the belief that conservation easements can be a blunt approach to a very dynamic need ... The permanency of conservation easements can go against the need for being dynamic because of the changing nature of agriculture, climate, human nature, markets, business ... you know, everything else.

And while the term "farmland commons" implies multiple farmers and farm families living and farming on the same property, McSweeney notes that many farmland commons organizations support only one farmer. In this way they function similarly to BPS programs, although without the eventual transition to conventional, individual ownership. In most cases, the farmer is also expected to take an active role in the "commons" organization, McSweeney says.

> The farmland commons model allows the farmer and the board to have a long-term conversation and dialog about the stewardship of the farm. It's a model that requires people to work together and have some boundaries and expectations for stewardship and farm practices, which can be a better tool for farmland preservation than conservation easements in a lot of situations.

Although the Farmers Land Trust has worked with specific communities to establish "commons" farms, their broader mission is to promote the farmland commons model across the United States and provide resources for local groups seeking assistance and resources for establishing local farmland commons projects. After a career in social work and real estate, and a long volunteer history in sustainable food system development, McSweeney formed the Farmers Land Trust with his partner and co-executive director, Kristina Villa, as an open-source resource to help others create their own commons structures. "Kristina and I saw the interest and opportunity and need for this model … so we formed the Farmers Land Trust to directly focus on this farmland commons work," he said.

Officially formed in 2022, the organization has a diverse and growing staff as of 2025, including a writer, copy editor, and photographer to help promote the work of farmland commons across the country. McSweeney and Villa are frequent consultants and speakers on farmland commons initiatives, and the organization's website serves as a resource library for various organizational and legal document templates necessary for forming a local farmland commons organizations and projects.

If fostering land access for small-scale and beginning farmers and promoting sustainable agriculture practices are among the primary benefits of the farmland commons, McSweeney also highlights the role they can play in fostering a local agriculture economy and promoting the role of "local" agriculture. Creating partnerships with local land trusts and conservation groups is part of that community-building effort. Transitioning long-standing farms from a senior generation to new ownership is another. McSweeney says:

> Whether it's farmers in need of land who have been excluded from land ownership; whether it's long-standing organic farms that want to carry forward into the next generation; whether it's nonprofit organizations coming together because they value community food security … the farmland commons is really a model that can be used for holding real

> estate that is important for the community. It's about how a community can secure their land base for whatever their local values are. So it's not just land conservation or protection. It's supporting human community and transitioning farms into the next generation.[14]

Land Tenure and the Politics of Place

In his scholarly essay on farmland commons, legal scholar Jamie Baxter emphasizes the ripple effect farmland access and tenure issues play on community land use and food production, "thereby reducing farmers' incentives and capacities to invest in practices and material improvements that would achieve better ecological, economic, and social outcomes."[15]

Like Garrett Hardin, who described the challenges of collective ownership models in his famous article "The Tragedy of the Commons," which argues that individuals acting in their own self-interest will ultimately deplete shared resources, harming everyone within the "collective" or community, Baxter recognizes that commons ownership models are no panacea for elevating the sustainable agriculture movement. However, it can still serve as a pragmatic, local solution to the land access and tenure insecurity that plague many local farming communities in America, he argues.

In the same way, the skeptic can argue that BPS solutions like that of the Ann Arbor Greenbelt Program or the Conservation Fund's "Farms Fund" initiative can never achieve the scale necessary to successfully combat the rapid farmland conversion occurring in Ann Arbor or on the metropolitan outskirts of American cities like Atlanta, Georgia. But Baxter argues that these "micro" approaches can create positive cumulative effects, creating pathways for broader community engagement between local farmers and others within their local communities. "When local institutions become the focal point for policy development in the area of farmland succession and tenure, small farmers can utilize their comparative advantages to participate effectively in local governance and the processes of decision-making," Baxter writes.[16]

It is an argument Kemmis would not only recognize but also applaud. For despite their limited scope, BPS and farmland commons approaches engender the type of cooperation and common inhabiting Kemmis believes necessary for engaged citizenship and the politics of place. They also engender communal values—that farmland preservation matters in these places, but also that farmland preservation can be a component of a broader effort to sustain the local farming system. "As rural areas transition, meanings and identities become contested and struggles over land use and landscapes become ways of deciding what does and does not belong there," write Jeffrey Masuda and Theresa Garvin. "In rural areas, perhaps more than elsewhere, place is more than just the value of the land. Rather place is a struggle over identity and power to impose one meaning over another."[17] If true, "new" tools like "Buy-Protect-Sell" and a modern iteration of the farmland commons model impose a meaning of preservation and stewardship in the places where they operate that update the farmland preservation ethos while adapting it for the issues of this generation.

Notes

1 "Status of State Pace Programs 2025," AFT Farmland Information Center, accessed November 4, 2025, https://farmlandinfo.org/wp-content/uploads/sites/2/2024/10/status-state-PACE-programs-2025.pdf.
2 Gabrielle Roesch-McNally (Women for the Land Director) in discussion with the author, April 28, 2025.
3 Jennifer Dempsey (Farmland Information Center Director) in discussion with the author, March 27, 2025.
4 Holly Rippon-Butler in discussion with the author, May 7, 2025.
5 David Howard (National Young Farmers Coalition Policy Director) in discussion with the author, April 18, 2025.
6 Liz Carlisle, Maywa Montenegro de Wit, Marcia S. DeLonge, Adam Calo, Christy Getz, Joanna Ory, Katherine Munden-Dixon, et al. "Securing the Future of US Agriculture: The Case for Investing in New Entry Sustainable Farmers," *Elementa (Washington, D.C.)* 7, no. 1 (2019), doi:10.1525/elementa.356.
7 City of Ann Arbor, "Greenbelt Program Strategic Plan" (2019 revision), https://www.a2gov.org/greenbelt/.
8 Remy Long and Rosie Pahl Donaldson (Ann Arbor Greenbelt Program) in discussion with the author, April 30, 2025.
9 Margaret Ann Snow, "Love Is Love Cooperative Farm Profile, Growing For Market," July 31, 2023, https://growingformarket.com/articles/love-love-cooperative-farm-profile.
10 "Working Farms," The Conservation Fund, accessed July 22, 2025, https://www.conservationfund.org/our-priorities/working-lands/working-farms/.
11 Krisztián Varsa (Director, Farms Fund) in discussion with the author, May 8, 2025.
12 Allan Greer, "Commons and Enclosure in the Colonization of North America," *The American Historical Review* 117, no. 2 (April 2012): 365–386.
13 Lynn Pitman, "History of Cooperatives in the United States," University of Wisconsin Center for Cooperatives (2018).
14 Ian McSweeney (The Farmers Land Trust Co-Executive Director) in discussion with the author, May 23, 2025.
15 Jamie Baxter, "Recovering Farmland Commons," in *Sustainable Food Futures*, eds. Jessica Duncan and Megan Bailey (Routledge, 2017), 61.
16 Jamie Baxter, "Legal Institutions of Farmland Succession: Implications for Sustainable Food Systems," *Maine Law Review* 65, no. 2 (Jan. 2013): 408.
17 Jeffrey R. Masuda and Theresa Garvin, "Whose Heartland?: The Politics of Place in a Rural-Urban Interface," *Journal of Rural Studies* 24, no. 1 (2008): 115.

Bibliography

American Farmland Trust Farmland Information Center. "Status of State Pace Programs 2025." https://farmlandinfo.org/wp-content/uploads/sites/2/2024/10/status-state-PACE-programs-2025.pdf.

Ann Arbor, City of. 2019. "Greenbelt Program Strategic Plan." https://www.a2gov.org/greenbelt/.

Baxter, Jamie. 2013. "Legal Institutions of Farmland Succession: Implications for Sustainable Food Systems," *Maine Law Review* 65, no. 2: 408.

Baxter, Jamie. 2017. "Recovering Farmland Commons." In *Sustainable Food Futures*, edited by Jessica Duncan and Megan Bailey. Routledge.

Carlisle, Liz, Maywa Montenegro de Wit, Marcia S. DeLonge, Adam Calo, Christy Getz, Joanna Ory, KatherineMunden-Dixon, *et al.*2019. "Securing the Future of US Agriculture: The Case for Investing in New Entry Sustainable Farmers," *Elementa (Washington, D.C.)* 7, no. 1 (2019). doi:10.1525/elementa.356.

Greer, Allan. 2012. "Commons and Enclosure in the Colonization of North America," *The American Historical Review* 117, no. 2: 365–386. doi:10.1086/ahr.117.2.365.

Masuda, Jeffrey R., and Theresa Garvin. 2008. "Whose Heartland? The Politics of Place in a Rural-Urban Interface," *Journal of Rural Studies* 24, no. 1. doi:10.1016/j.jrurstud.2007.08.003.

Pitman, Lynn. 2018. "History of Cooperatives in the United States." University of Wisconsin Center for Cooperatives. https://resources.uwcc.wisc.edu/History_of_Cooperatives.pdf.

14 Bridging Organizations and Transformational Leadership

"Places have a way of claiming people. When they claim very diverse kinds of people, then those people must eventually learn to live with each other; they must learn to inhabit their place together," Kemmis writes in *Community and the Politics of Place*. "It is through the nurturing of such virtues (and in no other way) that we might begin to reclaim the competency upon which democratic citizenship depends."[1]

Throughout this book, I've written about how *Community and the Politics of Places* influences my thinking about farmland preservation. As the scholar Jake P. Greear writes, Kemmis's book

> shows how, by reenlisting our economics and our politics in the service of small-scale, self-governed politics, the populist impulse and environmental sensibilities can become, must become … allied with one another. This reexamination of American politics, and by extension the politics of advanced industrial societies everywhere, offers the social and political critique needed to give new direction … into human ecological interaction."[2]

Greear's observations on small-scale civic participation still ring true, but only a simpleton would argue the world is the same now as it was in 1990 when *Community and the Politics of Place* was published. And some would argue that the idea of place has become more abstract, more virtual, since then, complicating our notions of how and where civic participation takes place.

But that abstraction only makes the nurturing of place virtues and social capital more important than ever before. If social capital refers to the collective power embedded in networks of people who live and work in a particular place or community, and that power can be oriented around place-based initiatives, such as the need to preserve farmland or rural character, the question becomes how we can organize or capture social capital to initiate meaningful change in an increasingly digital world.

One answer to that question is the idea of "bridging" organizations. Recall that bridging social capital (or simply "bridging capital") is the type of social capital that develops through outward-looking social networks that cross

DOI: 10.4324/9781003644545-17

social divides and backgrounds. Research indicates that bridging capital can increase access to power or capital, increase tolerance and acceptance of different values, and increase acceptance of new ideas and approaches, among other social benefits. It is often contrasted with bonding social capital (or "bonding capital"), which is the form of social capital that develops within groups and focuses on characteristics such as trust, conflict, and agency in smaller, homogenous social networks that have developed over time such as families, churches, and boards of directors.[3]

As their name suggests, bridging organizations help link local groups and individuals across scales and hierarchies of society and government. Although bridging organizations typically work within government bureaucracies, they are usually not government agencies themselves but, like land trusts, are nonprofits or other forms of nongovernmental organizations (NGOs). This "extra-governmental" characteristic allows bridging organizations to operate both inside and outside of government, with more flexibility to serve as facilitators, innovators, and leaders, avoiding many of the roadblocks imposed by government bureaucracies and regulations. "Bridging organizations can be described as providing social incentives to stakeholders to invest in building trust, identification of common interest, and resolving conflict," writes the environmental policy scholar Carl Folke and his research colleagues.[4]

Researcher Judy Stewart defined four characteristics of bridging organizations: (1) they have formalized organization with their own resources and personnel; (2) they have network connections across scales and hierarchies of government and community; (3) their objectives are strategic; and (4) they provide a forum for learning, knowledge, trust-building, and conflict resolution by acting as facilitators and attracting expertise and knowledge.[5] For Folke and his co-authors, this last characteristic may be the most important one. "The facilitation, leadership, and social incentives for collaboration provided by bridging organizations … appear to be essential for building the capacity to change," they write.[6]

In farmland preservation, this bridging role is frequently played by land trusts, who are able to use their organizational resources to connect farmers and agricultural landowners throughout the broader conservation and government agency community. Although each land trust is obviously different, many land trusts, in particular larger ones, possess all four of Stewart's defining characteristics. For example, most land trusts are governed by boards comprised of prominent community members with deep and high-level relationships both within the communities they serve and beyond them. Land trusts' strategic conservation objectives are (usually) strategically aligned with the farmland preservation movement, and most possess a high level of technical expertise in tools such as conservation easements and PACE programs.

Some land trusts have significant financial resources, or access to these resources through donors and philanthropic supporters. But because most are supported to some degree by grants and government programs, they develop entrepreneurial capacity to adapt their conservation goals to these available funding sources, leading to flexible and creative preservation approaches. This entrepreneurial character also helps land trusts establish connections with a wide variety of public agencies, NGOs, and philanthropic funders both within and beyond the communities they serve, creating the type of bridging capital that attracts outside expertise and resources but also facilitates problem-solving and sustained, intentional interactions that build trust and social capital up and down its scales of interactions, from the individual landowner to the halls of Congress.

This type of extra-governmental network is sometimes referred to as a "shadow network," which operates both within and outside of official government channels, creating incubation for new governance approaches. Shadow networks feature a "willingness to experiment and generate alternative solutions to emerging problems," writes the researcher Per Olsson and his research colleagues, who believe that the emergence of shadow networks is typically triggered by a social or ecological crisis. Because of their inherent flexibility and network characteristics, shadow networks are better able to develop entrepreneurial approaches to these crisis situations than traditional government, and they are often adept at equipping themselves with innovative resources, including the identification and mobilization of economic incentives for change. Shadow networks are also frequently more adept than conventional government at attracting social capital to address wicked problems. "Shadow networks can provide a platform/arena for collaboration in which this social capital can be reorganized and recombined; this in turn can generate innovation and nurture renewal in times of reorganization."[7]

The Athens Land Trust as a Bridging Organization

Few land trusts provide more bridging functions across the issues of land preservation, community access and equity, and even affordable housing and community revitalization than the Athens Land Trust (ALT) in Athens, Georgia. Although the organization has preserved over twenty thousand acres, its vision has always been a broader one focused on a community ethic that puts community above resource protection.

That vision belongs to Nancy Stangle, who founded ALT with her colleague Skipper Stipe Maas back in 1994. At the time, the two women were working to establish a communal residential community on 139 acres near Athens. "We wanted it to be an intentional community but not one with a lot of dogma. We were just wanting to be in community and live as lightly on the land as possible," recalls Stangle. Their plans were based on a European co-housing model, combining private dwellings with shared resources and common space, similar to the farmland commons model discussed in the previous chapter.

Through this process, Stangle learned about the mechanics of land preservation and conservation easements and how "community land trusts" were working to revitalize minority neighborhoods with affordable housing initiatives in the Atlanta metro area. A community land trust is a nonprofit organization that adopts the commons model, acquiring and holding land in trust to ensure long-term affordability. Housing is typically offered through ground leases, which allow residents to own their homes but not the land where they are built, keeping land accessible for low- and moderate-income residents while providing community stewardship and oversight. "So it just hit me—like maybe we are supposed to start a land trust that does both, land conservation and affordable housing. To me, it was more of a holistic view," Stangle remembers.

Stangle's "revelation" led to the creation of one of the most adaptive and community-oriented land trusts in the nation. Today, ALT's work focuses not only on the intersection between land conservation and affordable housing and community, but also on building sustainable food systems and, more recently, addressing the farmland transition and agricultural viability crises. "The early origins were combining the community land trust and conservation land trust models," agrees Joe Barbaree, ALT's current director of land conservation. "But the third leg of that stool was the agricultural piece, which was also there in the beginning."

At first the organization focused on providing access to common farm and garden spaces as part of its community development work, Barbaree says. "A lot of our work early on was really grassroots, really doing a lot of community organizing and understanding the needs of community. And building up the education and empowerment side of our work." During the interview, Barbaree points out that he is speaking from an office located withing the ALT's incubator farm facility, the 5-acre Williams Farm, located just 1.5 miles outside of Athens. At this facility, ALT offers a two-year apprentice program for beginning farmers from minority or veteran backgrounds.

But as the organization's capacity and expertise grew, it also expanded into the arena of traditional farmland preservation through conservation easements. This shift represents the adaptation and entrepreneurial spirit that is typical of bridging organizations. "From a conservation perspective, I think our emphasis early on became farmland and working forests," explains Barbaree, who credits the acceleration of their farmland preservation work to two factors. The first was the Natural Resources Conservation Service's (NRCS) working lands conservation easement acquisition program, now known as the Agricultural Conservation Easement Program (ACEP), which was established in the late 1990s but became more widely utilized by land trusts in the 2000s. The second was the Land Trust Alliance's Accreditation Program, which started in 2006. For years, explains Barbaree, ALT was the only accredited land trust in Georgia. "So that gave a boost to our farmland and working farms conservation work."

Today, ALT can boast about the permanent protection of over 16,000 of working farmland and forests.[8] But these days the organization is as well known for its community food system work as its traditional farmland preservation. The Williams Farm campus is a hub for much of this work, hosting workshop series on business planning and sustainable farming techniques in addition to its incubator program. ALT also offers a Young Urban Farmers program, where high school students from disadvantaged backgrounds learn technical gardening and farming practices and entrepreneurial skills for developing their own small-scale agricultural businesses.

Working with youth has always been a focus of ALT's approach, a lesson Stangle learned through her focus on community. "Many of these neighborhoods were desperate to get their youth more engaged," she recalls. Again, ALT adapted to meet the needs of its constituents. "All of our programs have a youth component now. This came out of the community saying, 'We've got to have opportunities for youth because otherwise we're going to lose them.'"

In addition to the Young Urban Farmers program, ALT also runs a Young Urban Builders program through its affordable housing initiative. Established in 2018, the program is designed to provide job training and experience in the construction and home repair industry while assisting directly in neighborhood revitalization projects. Another recent youth initiative is ALT's Young Conservation Steward program, a summer program formed in conjunction with the Northeast Georgia Invasive Plant cooperative, where participants learn important environmental skills like invasive plant removal and habitat restoration.[9] Stangle's instincts about working with youth are supported in recent scholarly research suggesting that youth programs can bring a greater sense of purpose and cohesion to community engagement initiatives, enhancing social capital.[10] In return, youth participation in community programming enhances their sense of efficacy and self-esteem, empowerment, open-mindedness, and personal responsibility.[11]

For Stangle, ALT's farm preservation work has always been rooted in its overall community development and revitalization objectives, reflected in its grassroots programming. "We got into doing community gardens in an interesting way because we were working in these historically low-income, or low-wealth, neighborhoods," she recalls.

> Our community garden work actually started from a neighborhood revitalization perspective ... We've established community gardens in apartment complexes, mobile home parks, senior living facilities, and revitalizing neighborhoods ... So to me, there was a bridge between our farmland protection work, our neighborhood revitalization work, and our mission toward empowerment and what we were doing with affordable housing.

"Even if we're not preserving land directly, we try to do projects that preserve rural spaces that benefit the community," agrees Barbaree. With their focus on community, ALT has always specialized in connecting local communities with outside resources, a classic hallmark of bridging organizations. "Linking different actor groups in networks and creating opportunities for new interactions are important for dealing with uncertainty and change and critical factors for learning and nurturing integrated adaptive responses to change," observe Folke and his colleagues. One of the key benefits of this bridging process is social learning and agency, which are essential both for the cooperation of partners and successful outcomes for these partnerships. Social learning is the collective learning that occurs through iterative practice, evaluation, and modification. Agency is the way individuals or groups use power or capacity to act in purposeful and meaningful ways.[12] Bridging organizations provide a forum for interaction of these different kinds of knowledge, and the coordination of other tasks that enable cooperation, observes scholar Fikret Berkes.[13]

Indeed, a brief conversation with Stangle reveals a plethora of local and federal agencies that ALT has connected to local communities—planning and zoning officials, the University of Georgia, Georgia Cooperative Extension, the U.S. Department of Housing and Urban Development (HUD), the Athens Grow Green Coalition, United States Department of Agriculture (USDA) and NRCS, the Georgia Heirs Property Law Center, and Land Trust Alliance, to name just a few. "I have to say one of the things I'm most proud about is how much collaboration we've had over the years," adds Stangle. "I mean, we haven't done any of these programs by ourselves."

This bridging function is where Barbaree believes the future of ALT's farmland preservation work lies. "On the landowner side, the growth potential for conservation work is more and more of the resources going to landowners," says Barbaree.

> We've done a good amount of that work over the years. But I think more and more, we want to serve as that clearing house for connecting landowners to information and resources and do the best job we can at that.

Barbaree points specifically to stewardship and management expertise for landowners as a place where the Athens Land Trust can serve as a bridging resource for their conservation easement donors.

Despite ALT's innovations in broadening the mission and programs of the traditional land trust model, Stangle sees the continuum of community in all of the organization's initiatives. For ALT, the focus has never been on land protection, but on building community and connecting it to resources. "It comes down to listening to what people tells us they need," Stangle says. This community-focused commitment embodies Kemmis's "second language of commitment"—the way our language and interactions with others acknowledge the shared responsibilities of place. "So, we've adapted and tried to see how we as an organization can help solve those problems."[14]

Transformational Leadership and the California Agricultural Leadership Foundation

Much of the research literature that emphasizes the social capital benefits of bridging organizations in natural resource policy also emphasizes the role of leadership. "Leadership is essential in shaping change and reorganization," write Folke and his colleagues. "Leaders can provide key functions ... such as building trust, making sense, managing conflict, linking actors, initiating partnerships among actor groups, compiling and generating knowledge, and mobilizing broad support for change."[15]

"Transformational leadership" is a label frequently used to define this brand of visionary, collaborative leadership. Transformational leadership is leadership that facilitates the achievement of goals through heightening consciousness of collective interest. Transformational leaders often possess high levels of expertise and familiarity with technical aspects of problems yet focus on addressing these issues through a shared-goals orientation that achieves bridging connections across networks and hierarchies of social groups.[16]

Through her years of dedication to community and building trust, her visionary commitment to farmland preservation and the expertise she developed in its tools and processes, and her success at linking her constituents with outside agencies and resources, Nancy Stangle is a textbook definition of the transformational leader. But such leadership is rare and increasingly difficult to find in many of America's fragmenting rural communities. Organizations like the California Agricultural Leadership Foundation (CAFL) address this scarcity by training emerging agricultural leaders in the critical skills necessary for tackling the complex challenges facing today's agricultural industry and its communities. "The California Agricultural Leadership Foundation was initially founded with the intention of building leaders for California agriculture who otherwise maybe wouldn't have those opportunities," explains Dane White, the CAFL's director of education. "If you go back to the origin of this organization, this whole thing has been about how do we build capacity to wield influences in spaces where it's needed for our farmers and agricultural communities."

First founded in 1962, CAFL now has over 1,400 graduates, who are selected based on a competitive application process for twenty-four spots in a seventeen-month curriculum. These "fellows" train in a rigorous leadership curriculum that includes leadership theory, critical thinking, change management, and emotional intelligence, among other leadership skills and tools. The curriculum is supported by the CAFL's five partner universities, including Cal Poly Pomona, Cal Poly San Luis Obispo, Fresno State, UC Davis, and Chico State. "The curriculum is really based on increasing personal power and influence through the development of self and the development of skill," explains White. The program focuses on building the fellows' emotional intelligence through psychometric assessments, seminar trainings, and one-on-one work with an executive coach. "So we're really trying to elevate their authenticity, their empathy, and their effectiveness, which are our three primary program outcomes."

While some leadership programs focus on advocacy skills, the CAFL aims for a more holistic approach that fosters deeper relationships and influences—that is, building the type of social capital necessary for the politics of place. One of the ways it does this is through exposing fellows to new perspectives and approaches outside of California. "Our fellows do a national seminar, so we'll go to Washington, D.C., and then another state to learn about their social systems and systems of influence. And then we also do an international seminar." In 2025, fellows took a two-week trip to study agriculture in South Africa and Zambia. "It's all built on the leadership theory that your capacity magnifies when you learn the world does not operate based on your own personal comfort," White explains, embodying concepts of bridging social capital.

These trips help establish the CAFL's focus on empathy, which White describes as the ability to understand someone else's perspective and lived truth is valid.

> When you've got leaders who can come into a room with a group of people with the ability to listen, to communicate, to ask good questions and be curious ... Then I think you're creating leaders who can have some influence constructively. They are seen as more trustworthy and [the] type of people that others want to listen to.

This empathy focus helps develop agricultural leaders who can engage in conversations allowing for greater bridge-building and constructive solution creation in their communities, defining characteristics of transformational leadership. It's also a great description of Kemmis's "second language of commitment." Indeed, the program emphasizes the importance of local leadership and engagement in a way that would make Kemmis proud. "One of the things we ask our fellows to articulate very specifically as part of this training is how they're going to make an impact in their community," says White. "We really do ask them to take a specific focus on building influence as a leader in a place where they can make a difference."[17]

Grow Food Carolina and Bridging for Food System Development

If the bridging organization role of the Athens Land Trust tends to focus on linking its community constituents vertically up and down different hierarchies and levels of government, other agricultural bridging organizations are focused on connecting farmers horizontally across customers, vendors, and suppliers, supporting farmland preservation through agricultural viability and the development of local food system economies.

GrowFood Carolina is this kind of bridging organization. According to its website, its mission is to foster the assets of the Charleston, South Carolina, region's small-scale agricultural producers "to help create a stronger rural economy, spur job creation, and build capacity in rural communities by connecting farm businesses to the thriving local food movement."[18] The organization was founded in 2011 as a "food hub" warehouse in downtown Charleston. According to the National Food Hub Collaboration, a partnership of USDA's Agricultural Marketing Service, a food hub is a "business or organization that actively manages the aggregation, distribution, and marketing of source-identified food products primarily from local and regional producers to strengthen their ability to satisfy wholesale, retail, and institutional demand."

Food hubs serve several important functions: they increase market access for local and regional producers; add value to current food distribution systems; and demonstrate innovative business models for small- and mid-scale agricultural producers.[19] Kemmis embraces this same idea in his discussion of the "city-state," the city center and its surrounding rural areas, which he believes comprises a natural system for both local politics and economics. If these cities and their regions can establish local economies, they can replace the isolation and exploitation of global economies with ones that instead support local social capital and civic participation, he argues in *Community and the Politics of Place*.[20] These linking processes also embody the role of bridging organizations, which build social capital through enhancing networks, knowledge, capacity, and trust within social-ecological systems.

But theoretical arguments aside, the main role of GrowFood Carolina is helping small-scale agricultural producers with the marketing and distribution services they need to compete against the industrial agriculture segment, says Benton Montgomery, GrowFood Carolina's current director.

> Having a team like GrowFood Carolina with resources and skilled staff to walk our farmers and producers through that last part of the process—the distribution and marketing part—is immensely helpful for our farmers. It gives them more time to actually tend to their crops.

GrowFood builds this capacity through a variety of services and resources. At their warehouse facility in Charleston, they have cold storage, a vital but expensive resource for many small- and medium-scale agricultural producers. GrowFood also operates a fleet of refrigerated delivery trucks. "Our farmers don't have to put that huge investment into refrigeration on site, as well as having refrigerated vehicles. So, we take those initial capital costs off their hands," Montgomery says.

On the marketing side, GrowFood has built established networks within Charleston's thriving restaurant industry, another example of its bridging functions.

> There's a huge culinary demand for local foods in Charleston, as far as the dedicated restauranteurs and chefs who want to be supportive of their neighbors and the local food market. They like the personal touch

> they're able to have with us and with the farmers versus their traditional distributors, where they have zero access to the farmer.

Montgomery says local chefs often give them feedback on the type of local agricultural product they are seeking, which the organization then passes back to farmers. "It allows the chefs to have a much closer touch with the finished product they are producing for consumers." The relationship also works reciprocally. "Our relationships allow us to tell the chefs, 'Okay, this is what's coming in. We're going to have a lot of this,' which helps them plan their menus as well."

Montgomery describes GrowFood's work with chefs and specialty retailers as the "first tier" of their operational model. The second tier is their Community Supported Agriculture (CSA) program, linking local farmers directly with consumers through a weekly subscription service during the growing and harvest season. This service started in 2020. "That's also when we developed our food access side," says Montgomery, which relies on individual fundraising to supply fresh local food for disadvantaged members of the community. "We started raising money to purchase product from our suppliers and then send it to food banks and other community partners for areas that needed fresh food during the pandemic."

Although grown out of the Covid crisis, Montgomery says the food access program has turned into a big part of GrowFood's work, now comprising the third tier of their mission. The organization works with the Charleston County public library system to stock their "Free and Fresh" fridges, where anyone can take fresh and free local produce. They also work with thirteen different Head Start Centers in Charleston County to build fresh food boxes for between seven hundred and nine hundred families every other week. "The farmers really like knowing their produce is going to feed their neighbors," Montgomery says. "That kind of community aspect is really taking off for us."

Montgomery describes the organization's fourth tier as their grower service initiative, where they help their producers with crop planning for the upcoming year. "So, before each season we'll create a spreadsheet with them to help plan out the year. It tells them things like, 'Hey, if you bring us ten boxes of squash between these six weeks, you're going to receive an average of $15 back per case ... That's something the growers really appreciate and allows us to tell the chefs, 'Okay, this is what's going to be coming in.' It helps them plan out their menus."

Although the links between CSAs, specialty markets, business planning, and farmland preservation may be obscure, Montgomery sees them clearly in his day-to-day work. "Anything we can do to help our farmers save money or put money back in their pocket and allow them more time to actually maximize the use of their land can hopefully keep the land in agricultural production and keep farming on it," he explains.

This emphasis on farm-business viability and supporting local food economies is a more holistic approach to farmland preservation, but it is one

Kemmis would clearly recognize as a way local markets and economic systems spur the politics of "practice," a term he uses to describe the process of arriving at the "common by abstracting from the particular ... through a shared way of arranging and relating to the actual, physical world." In contrast, the rules and laws of the resource republic are abstractions, their purposes related to theoretical principles, not the real-world interactions and needs of community. "In this way places breed cooperation ... a politics of citizens working out problems and possibilities of their place directly among themselves implies a revival of the old republican notion of citizenship base upon civic virtue."[21]

In our interview, Montgomery doesn't specifically link the work of GrowFood Carolina to a resurrection of civic virtue, but he has seen firsthand how it fosters a sense of community among his agricultural suppliers and customers, along with one of community's central ingredients—trust. "I must say we're very strong at relationship-building. We're very high-touch with both our customers and our farmers. and being able to develop that trust is really vital to our program."[22] Montgomery doesn't refer to GrowFood Carolina as a "bridging organization" either, but through supporting local farmers and helping them find the markets for their local goods, they are building the bridging capital necessary for local farmers and food systems to thrive.

Notes

1 Daniel Kemmis, *Community and the Politics of Place* (University of Oklahoma Press, 1990), 118–119.
2 Jake Greear, "Toward a Political Rehabilitation of Environmentalism: Daniel Kemmis's Community and the Politics of Place," *Organization & Environment* 18, no. 3 (2005): 339.
3 H. Coffé and B. Geys, "Toward an Empirical Characterization of Bridging and Bonding Social Capital," *Nonprofit and Voluntary Sector Quarterly* 36, no. 1 (2007): 121–139.
4 Carl Folke, Thomas Hahn, Per Olsson, and Jon Norberg, "Adaptive Governance of Social Ecological Systems," *Annual Review of Environmental Resources* 30 (2005): 462.
5 Judy Stewart, "Critical Roles of Bridging Organizations in River Basin Management in Alberta, Canada," *WIT Transactions on Ecology and the Environment* 234 (2019): 91–101.
6 Folke et al., "Adaptive Governance," 462.
7 Per Olsson et al., "Shooting the Rapids: Navigating Transitions to Adaptive Governance of Social-Ecological Systems," *Ecology & Society* 11, no. 1 (2006): 12, 13.
8 "The Numbers," Athens Land Trust, accessed August 2, 2025, https://athenslandtrust.org/about/our-impact/.
9 "Youth Development," Athens Land Trust, accessed August 4, 2025, https://athenslandtrust.org/our-work/education/youth-development/.
10 N.T. Wong, M.A. Zimmberman, and E.A. Parker, "A Typology of Youth Participation and Empowerment for Child and Adolescent Health Promotion," *American Journal of Community Psychology* 46, no. 1 (2010): 100–114.

11 B. Checkoway, K. Potuckuchi, and J. Finn, "Youth Participation in Community Planning: What Are the Benefits?" *Journal of Planning Education and Research* 14, no. 2 (1995): 134–139.
12 Folke et al., "Adaptive Governance," 441, 462.
13 Fikret Berkes, "Evolution of Co-Management: Role of Knowledge Generation, Bridging Organizations, and Social Learning," *Journal of Environmental Management* 90, no. 5 (2009): 1692.
14 Nancy Stangle and Nigel Strong (Athens Land Trust founder and staff) in conversation with the author, April 8, 2025, and April 25, 2025, respectively.
15 Folke et al., "Adaptive Governance," 451.
16 Nicole Adroin, Rachelle K. Gould, Elin Kelsey, and Priya Fielding-Singh, "Collaborative and Transformational Leadership in the Environmental Realm," *Journal of Environmental Policy and Planning* 17, no. 3 (2014): 360–380.
17 Dane White (California Agricultural Leadership Foundation director of education) in conversation with the author, April 17, 2025.
18 "GrowFood Carolina," Coastal Conservation League, accessed August 5, 2025, https://coastalconservationleague.org/projects/GrowFood/.
19 James Barham, Debra Troop, and Kathleen Enterline, "Regional Food Hub Resource Guide," U.S. Department of Agricultural Marketing Service, April 2012, https://ageconsearch.umn.edu/record/145227/?v=pdf, quote on p. 4.
20 Kemmis, *Community and the Politics of Place*, 84–108.
21 Kemmis, *Community and the Politics of Place*, 122–123.
22 Benton Montgomery (GrowFood Carolina executive director) in conversation with the author, May 9, 2025.

Bibliography

Adroin, Nicole, Rachelle K. Gould, Elin Kelsey, and Priya Fielding-Singh. 2014. "Collaborative and Transformational Leadership in the Environmental Realm," *Journal of Environmental Policy and Planning* 17, no. 3. doi:10.1080/1523908x.2014.954075.

Barham, James, Debra Troop, and Kathleen Enterline. 2012. "Regional Food Hub Resource Guide," U.S. Department of Agricultural Marketing Service. https://ageconsearch.umn.edu/record/145227/?v=pdf.

Berkes, Fikret. 2009. "Evolution of Co-Management: Role of Knowledge Generation, Bridging Organizations, and Social Learning," *Journal of Environmental Management* 90, no. 5. doi:10.1016/j.jenvman.2008.12.001.

Checkoway, B., K. Potuckuchi, and J. Finn. 1995. "Youth Participation in Community Planning: What Are the Benefits?" *Journal of Planning Education and Research* 14, no. 2. doi:10.1177/0739456x9501400206.

Coffé, H., and B. Geys. 2007. "Toward an Empirical Characterization of Bridging and Bonding Social Capital," *Nonprofit and Voluntary Sector Quarterly* 36, no. 1. doi:10.1177/0899764006293181.

Folke, Carl, Thomas Hahn, Per Olsson, and Jon Norberg. 2005. "Adaptive Governance of Social Ecological Systems," *Annual Review of Environmental Resources* 30. doi:10.1146/annurev.energy.30.050504.144511.

Greear, Jake. 2005. "Toward a Political Rehabilitation of Environmentalism: Daniel Kemmis's Community and the Politics of Place," *Organization & Environment* 18, no. 3. doi:10.1177/1086026605279464.

Kemmis, Daniel. 1990. *Community and the Politics of Place*. University of Oklahoma Press.

Olsson, Per, L.H. Gunderson, S.R. Carpenter, P. Ryan, L. Lebel, C. Folke, and C.S. Holling. 2006. "Shooting the Rapids: Navigating Transitions to Adaptive Governance of Social-Ecological Systems," *Ecology & Society* 11, no. 1. doi:10.5751/es-01595-110118.

Stewart, Judy. 2019. "Critical Roles of Bridging Organizations in River Basin Management in Alberta, Canada," *WIT Transactions on Ecology and the Environment* 234. doi:10.2495/rbm190101.

Wong, N.T., M.A. Zimmberman, and E.A. Parker. 2010. "A Typology of Youth Participation and Empowerment for Child and Adolescent Health Promotion," *American Journal of Community Psychology* 46, no. 1. doi:10.1007/s10464-10010-9330-0.

15 Narrative Networks

Farmland preservation in the politics of place is based on the underlying premise that local matters. American farms and farmland exist in no single type of community. So why should farmland preservation policy in rapidly growing regions like Charleston, South Carolina (or Chatham County, North Carolina, for that matter), resemble farmland preservation in the farming communities of Kansas or Wyoming?

When Kemmis describes the "politics of inhabitation," he writes that it is contingent on the human virtues and patterns of relationships dependent in place, not the procedures and bureaucracies of the resource republic. To solve wicked problems like farmland loss, he argues, we cannot simply label them as "the government's problem, nor can the government solve it by itself. The solution must be sought on the level of citizenship."[1]

Indeed, scholarly research affirms that these types of complex land use and environmental issues can only be solved through adaptive solutions born of local community knowledge and cooperation. Yes, government can play a role. Policy can play a role. But it is social capital and community centered in place that makes the difference, and the foundations of adaptive governance are built on networks of social capital grounded in local ecological knowledge, advocacy, and civic participation.

True, the world has changed fundamentally since Kemmis wrote *Community and the Politics of Place* in 1990. For better *and* for worse perhaps, but the sustainable farming ethos of the "New Agrarians" has endured and proven itself adept at adaption to the networking and communication technologies of our digital age. The American Farmland Trust's "Women for the Land" program, for example, relies heavily on the Internet to link women farmers across the United States, facilitating peer-to-peer learning and training on the unique issues they face in their local communities. The Young Farmers Network uses the same technologies for America's cadre of new and beginning farmers. These initiatives don't ignore the politics of place. Indeed, their mission is supporting the small- and medium-scale farmers whom both Jefferson and Kemmis agree are necessary for conducting it. But they recognize that the answers to today's wicked environmental and economic problems must be developed through broader networks of like-minded

DOI: 10.4324/9781003644545-18

advocates and constituents that function beyond place's physical boundaries. And they are using the power of narrative to build these networks.

Narrative Networks and Women for the Land

Researcher Raul Lejano and his collaborators, Mril and Helen Ingram, use the term "narrative networks" to explain how people understand their relationships through stories, and how stories can connect people, ideas, and actions across social and institutional contexts. Their research is based on the concept of environmental networks—how people understand the relationship between themselves and their ecosystem—and the way that stories, or narratives, bind people and ecosystems together through a sense of history, common ground, and a vision of the future. When people express concerns or interests in their biophysical environment, and their role in it, they are expressing their social and political values in the processes that shape and influence that environment, not simply their concerns over specific biophysical attributes of place, such as its farmland. "Narratives are essential in catalyzing and sustaining environmental networks, and enabling them to exert influence," they write.[2] "Stories, or narratives, create the glue that binds people together in networks, providing them with a sense of history, common ground, and future, thus enabling them to persist even in the context of resistance."[3]

Recent research on stakeholder process in environmental policy has brought focus to the role of narrative and storytelling in collaborative governance, creating new awareness about the role of storytelling in the politics of place. Structural inequity is a common theme of these narratives, particularly in reaction to government or market breakdowns that disproportionately affect a certain segment or type of community member. Thus, agency and self-empowerment, or the pursuit of agency, also figure prominently in the themes of narrative networks, as do the establishment of personal and collective identities. "It is easy enough to say that we are the story that we tell," observe Lejano and his colleagues. "And when we associate ourselves with others, we are associating … our own stories with their own into a coherent metanarrative."[4]

Narrative networks seek to operationalize these metanarratives through place-based or identity-based advocacy, recognizing that connections to place and group consciousness are often expressed in terms of narrative. American Farmland Trust's (AFT) "Women for the Land" program employs metanarrative about the challenges facing women farmers as a tool for training, advocacy, and fellowship.

According to Gabrielle Roesch-McNally, AFT's director for Women for the Land, the network first emerged out of AFT's farmland preservation mission. "A big impetus in the beginning days was to reach women non-operating landowners and engage them in conversations about conservation," she recalls. According to an AFT study last updated in 2020, women are principal or co-principal landlords on more than 50 percent of United States farmland. And women landlords make up a third of the ownership of land that is rented out by non-operating landlords—landowners who own farmland but don't work it themselves.

Working with local and regional groups, AFT began establishing "learning circles" for these women landowners, which Roesch-McNally describes as "peer-to-peer learning groups" that created structured and formatted meeting spaces for women involved in agriculture. Participants in these circles regularly gather to hear from each other and agricultural resource providers about programs and practices that support farming issues such as soil health, farm viability, or legacy planning for their farmland. "The idea was to create spaces that are less hierarchical, more about shared learning and really helping women see that they have a lot of knowledge themselves that they can share with one another," says Roesch-McNally.

Gradually, however, Women for the Land's metanarratives expanded to encompass the broader barriers facing female farmers, particularly those belonging to ethnic minorities or just starting their farm businesses. According to a 2024 report by AFT (co-authored by Roesch-McNally), women statistically comprise a majority of beginning farmers but face more obstacles to financing and programmatic resources than their male counterparts. On average, female-operated farms are about half the size of their male-run counterparts, leading to lower revenues and limiting access to federal farm subsidies and other support programs that tend to favor the industrial scale of America's conventional, male-dominated farming operations. Female farmers often bear more family-care burdens than their male counterparts, such as child and eldercare, and are more likely to face gender-based violence on the job. "A growing body of evidence demonstrates that farming is among the most gender-unequal occupations in the U.S.," the report states. "We must reckon with the reality of unequal economic outcomes and access to resources for women agricultural producers."[5]

These gender-based inequalities play prominent roles in Women for the Land's advocacy and policy work, both on Capitol Hill and at agencies like the United States Department of Agriculture (USDA). "We have done some advocacy on specific legislative pieces," says Roesch-McNally, describing a recent effort to include provisions for affordable childcare within efforts to reauthorize the Farm Bill, America's major legislation governing agriculture, although helping women get access to the programs and resources of USDA and similar agencies is the heart of this advocacy effort.

Perhaps a more significant, if implicit, emphasis, however, is on building social capital for America's female farmers. "A big part of what we've tried to do in Women for the Land is build a more national sort of cohesive sense of what we do together," says Roesch-McNally.

> We think of the learning circles as an effective way to build community, build confidence, and build a support network for women producers. We're using our programmatic work as a way to inform the community of practice who are interested in making change on the ground.

Roesch-McNally highlights AFT's work with the Black Family Land Trust as an example of this type of collaborative approach. In October 2023, AFT's Women for the Land program partnered with the Black Family Land Trust and North Carolina State University's Local Food Program in a training focused on connecting BIPOC (Black, Indigenous, and People of Color) farmers.[6]

Not surprisingly, however, given AFT's traditional land preservation and conservation focus, land access remains one of Women for the Land's most prominent metanarratives. "A lot of our energy is being put into that topic because the land access piece is the elephant in the room for a lot of people in our work," says Roesch-McNally. "It's a huge issue and there's no silver bullet."

Through the broader work of AFT, Women for the Land supports "Buy-Protect-Sell" strategies, partnering with their vast network of local and regional land trusts and supporting programming through AFT's National Agricultural Land Network, its virtual farmland preservation community. They have also started working with USDA on a program to specifically help women and other minority farmers access financing for acquiring farmland, although Roesch-McNally wasn't sure if that program would continue when we spoke in 2025 due to the shifting priorities of the Trump administration.

Beyond advocacy and programmatic work, however, Roesch-McNally believes the biggest benefit of Women for the Land is the learning circle model, bringing female agricultural practitioners together in common space, uniting them around common narrative, and using those narratives to build social capital and agency. For evidence, she points to a study that indicated 73 percent of female farmers who participate in learning circles take some sort of action as a result of their participation, such as engaging in a new farm practice, engaging with USDA, or engaging with other farm stakeholders.

> We think there's an opportunity to brings more connection to those communities through this type of learning circle approach ... It's like building partnership with the folks on the ground who are connected to different producers and creating opportunities for them to come together.[7]

Of the learning circles' success, Roesch-McNally says,

> I think there is something magical that happens there. Or if not magical, it is intentional. Something really special happens in the learning circle that facilitates people making changes in their land that is different from what happens in other kinds of outreach events with farmers, particularly in terms of boosting people's confidence, boosting their connection to

> networks, and facilitating that feeling of being connected to other farmers and resource providers. So, I think I would argue the most important piece [of WFL] is fostering that sense of connection and building that network that has the ability to ripple out. It's like all these ripple effects of the connection that's happening there.[8]

Young Farmers Network

Like Women for the Land, the National Young Farmers Coalition (NYFC) adopts the tools of narrative networks to advocate for its member constituents—the emerging population of young and BIPOC farmers who are seeking training, assets, and institutional access for assistance in starting their own farming careers.

The issue is a critical one. As noted in Chapter 1, AFT calculates that more than 40 percent of American farmland is owned by farmers aged 65 or older, and USDA's 2022 "Census of Agriculture" reports that the average age of the American farmer is 58.1 years. Exacerbating the issue is the fact that America's oldest farmland owners tend to own and operate the small and mid-size farms that are most vulnerable to development, prompting AFT to estimate that 370 million acres of farmland is expected to change hands in the next twenty years as these farmers retire or die.[9]

Congress authorized a ten-member "Commission on Farm Transition" as part of the 2018 Farm Bill to address this issue. Unfortunately, however, the commission has not yet been constituted as of 2025. Meanwhile, the challenges facing younger farmers have only gotten more intense, according to David Howard, the policy development director for the NYFC. "There's a national crisis that's been unfolding for some time at the intersection of public policy, land consolidation, land use and prices, climate impacts, generational transition, and aging farmers," explains Howard. "And everything ties back to land access, because all of these challenges stack."

These challenges are magnified, Howard explains, because many of today's new and BIPOC farmers don't come from traditional farming backgrounds.

> Consistently you see that young farmers are not coming from a farming family, and that traditional model of taking over the family farm is less and less relevant to their experience. The farmers that we work with are coming to farming through a circuitous route. They're finding that passion somewhere along their journey and then leaning into it.

But that journey entails a plethora of challenges and barriers, which are encompassed in the metanarratives of NYFC. One of the biggest barriers facing new farmers is debt, Howard says, including student loans. "You know, this new generation of farmers is often financing things on credit cards, because the systems in place to access credit and land are inadequate to support their needs."

Access to financial capital, program support, and training are also part of this problem, as are auxiliary issues like affordable housing, childcare, and health insurance. But at the heart of the challenge narratives for young farmers are escalating land prices and access to farmland. The NYFC's 2022 survey ranked finding affordable land to buy as the top challenge for young farmers, with access to capital as the number two issue.[10] "And even if young farmers gain access, even if they're able to lease land, it's an uncertain relationship," adds Howard.

Holly Rippon Butler worked as the NYFC's land campaign director for ten years before starting her own consulting business focused on helping farmers and agricultural nonprofits navigate federal agriculture policy. During our interview, she described a host of issues keeping young farmers from accessing working farmland, ranging from rapidly escalating real estate prices, fueling land speculation, to the unique challenges young farmers face accessing federal loan programs.

> We see a lot of young farmers struggle to gain credit access, whether it's through the Farm Service Agency or local farm credit lenders. Because a lot of young farmers are running a really diversified farm operation, or maybe they're growing crops that the agents at the local offices just aren't familiar with, it's hard to fit into the paperwork of these traditional lenders.

Discrimination, difficulty accessing information and assistance, and the mismatch between business plans and land market conditions also exacerbate the access issue.

Sadly, the traditional conservation easement model for farmland preservation does little to help the land access issues for young farmers, Rippon Butler says. "The land trust model is not for everyone," she laments. "It can really be very exclusive."

Howard similarly complains that conservation easements do little to support the long-term viability of small- and medium-scale farms. "What conservation easements don't ensure is that a working farmer will own that land and be able to live off that land." He also worries about the "premium" conservation easements can put on property. "You can have land that's protected from development but still inaccessible to farmers because of the value of beautiful, protected land on the open market. People are going to be interested in that for their second home, or a retirement home or investment property."

Issues like these make farmland access a prominent metanarrative of NYFC's advocacy and outreach work across the country. In 2022, in a partnership with the Mexican food chain Chipotle, NYFC launched its "One Million Acres for the Future" campaign, which called for \$2.5 billion in federal funding to help new farmers access land. The campaign advocated for a dedicated fund that would help transition farmland from retiring farmers to young, beginning, and BIPOC farmers through tools such as cooperative ownership models and Buy-Protect-Sell programs. As part of the campaign, NYFC flew in one hundred young farmer leaders from across the country to Washington, D.C., in March 2023 for a series of over 150 meetings with leaders of Congress and USDA officials.

The effort seemed to find success when the Biden administration announced the USDA's "Increasing Land, Capital, and Market Access Program," whose purpose was to creatively address the interrelated challenges of access to land, capital, and markets for young and underserved producers by investing in community-based, locally led projects. "We saw this and we thought, 'Amazing, right? Like they get it,'" recalls Howard. "These things are all connected. We don't need one program here and another there. We need a program that sees the complexity of the issue and meets the people trying to solve it where they're at." Unfortunately, the program was frozen in January 2025, when the Trump administration took office, and in June 2025, USDA announced that the program would be cancelled.[11] However, NYFC is now advocated for the "New Producer Economic Security Act," a similar bill introduced in both the House and the Senate with bipartisan support in April 2025, which also focuses on community-led solutions to land access, capital, and market access issues (this proposed legislation remained under consideration as of the end of 2025 but had not been reported out of committee).

Given the agricultural policy priorities of the Trump administration, substantial new federal policy addressing agricultural land access is unlikely. But NYFC is still actively promoting these land access narratives on state policy levels. Howard and Rippon Butler collaborated on a report titled "Farmland for Farmers" that led to the State of New York allowing provisions in conservation easements that control resale prices.

The narratives of excluded and marginalized young farmers can be powerful tools for farmland preservation. NYFC uses stories of young farmers' struggle to access farmland as a tool for challenging the status quo of the United States agriculture system, raise awareness for the issues facing young farmers, and mobilize support for policy change. But it is at the local and regional level where NYFC's work addresses the deeper work of narrative networks, creating new modes of understanding and relationship for those who feel disenfranchised from entrenched power structures. One of its main tools for this local focus is NYFC's "Chapter Network," which the NYFC describes as "values-aligned groups of farmers working towards a common vision based on the needs of their communities." According to its website, NYFC supports chapters in twenty-five states, with some operating on a regional or local scale and some serving the entire state.[12]

"I think our chapters are full of farmers who are basically trying to find their way to a different food system," enthuses Howard, who had just spoken at a meeting of the NYFC's "Hudson Valley Young Farmers Coalition" (HVYFC) the night before our interview. "They are passionate, they are organizers, they are thinkers and doers who are coming up with solutions at the local level." Founded in 2012, the HVYFC describes itself as a "social and business network that encourages cooperation and communication among farmers" and boasts over 1,500 subscribers. According to their website, the chapter focuses on housing access and is "pursuing policy campaigns at the local, state, and federal levels to jumpstart responsible, farmer-led solutions to this crisis." As of summer 2025, they were also asking their members and other farmers in New York's Hudson Valley to share stories about impacts of USDA funding cuts mandated by the federal Department of Government Efficiency as part of the Donald Trump administration.[13]

To support this chapter work on a local level, NYFC operates a national organizing team as well as a "sub team" focused on what Howard calls "ladder engagement," or building farmer leaders. "I think that's basically what NYFC is really good at, being in community with farmers," says Howard. To support this "ladder engagement focus," NYFC has recently launched its "Root and Bloom Fellowship" program, an advocacy and leadership training opportunity for farmers across the country. Over nine months, twenty-five young farmers from across the country receive training in advocacy and power-building techniques in a series of virtual workshops. As a condition of participation, fellows are expected to implement an advocacy project in their own community and host meetings with local elected representatives on issues facing young and beginning farmers. They will also participate in supporting NYFC's active advocacy campaigns.[14]

For Rippon-Butler, the Root and Bloom Fellowship is only the latest iteration of NYFC's strengths in combining grassroots and national advocacy through engaging with narratives.

> I think NYFC is trying to elevate farmer voices, whether in the press, or in D.C. with members of Congress, or with land trusts and other organizations that we're working with in local communities. I think we got really good at advocating and building a big tent and working for progressive policy change in a way that brings policy makers from across the political spectrum to the table with a goal of working toward progressive change.[15]

Rural Advancement Foundation International and Trust in Narrative Networks

The Rural Advancement Foundation International also employs metanarratives about farmland access and the issues facing small- and medium-sized farmers. Known primarily by the acronym "RAFI," the organization's goal is to enhance access to resources and opportunities for historically underserved farmers, especially BIPOC farmers. They do this through what the organization calls "a flexible participatory model of community engagement" that includes training and advocacy efforts employing the metanarratives of farmland access, farmland tenure, and agricultural viability.

The organization offers a variety of ways to access their network. A hallmark of RAFI's programming is its Farm Advocacy Program, where farmers are connected with a trained farm advocate who then works one on one with farm families as a resource and voice for the often-confusing process of using federal farm support programs. These advocates also assist with a variety of other technical and financial issues, ranging from managing financial crisis to navigating federal farm and disaster programs to assistance in negotiations with lenders and creditors. For farmers in more immediate crisis, RAFI's Farmer Crisis Hotline connects them with qualified consultants and counsellors as quickly as possible. "The heart of what RAFI does, and really how we originated, is that direct work with farmers in crisis," explains Kavita Koppa, RAFI's executive director. "We've always focused on those people who are in between a rock and a hard place, and their farms are going to be lost in some form or another if they don't get some assistance."

A prominent example of narrative networks at RAFI is its "Farmers of Color Network" (FOCN), which provides technical assistance, training on funding opportunities and cost-share assistance, networking events, and other programming for farmers of color in the Southeast United States, lower mid-Atlantic, U.S. Virgin Islands, and Puerto Rico. "The Farmers of Color Network is exactly what it sounds like," explains Koppa. "It is a network focused on helping farmers of color not only get access to resources that they need but also really trying to hold a spot for people to connect one another and build their own forms of community."

According to Koppa, the FOCN includes over nine hundred BIPOC farmers across its service region. Although the FOCN often operates at a regional or virtual level, Koppa says the program originated out of a need to foster local political leadership. But the focus expanded as RAFI leaders realized the barriers that existed between their BIPOC constituents and local government infrastructures.

> What became very clear through that process was that there was a lot of distrust of those local decision-making bodies ... and also we were coming across more and more folks who were wanting to make their own thing. Instead of being a lone voice in local political arenas, they were more interested in a finding common voices for working together.

Lejano and his colleagues address this issue of trust in their writing on narrative networks. "Trust is said to be the instrumental element that allows sustained exchanges between members in the absences of contracts, regulatory mandates, or other conventional mechanisms," they write.[16] Trust is also a central component or "generator" of social capital, according to Putnam, Coleman, and other social capital researchers.[17] In narrative networks, relational aspects of networks are prioritized over transactional ones. "When people share a purpose or outlook or enemy, a narrative affirmation of the other members as trustworthy is extended to the extent that affirms one's own identity," write Lejano et al. "One can share one's own beliefs with kindred spirits." Collective identity generated through these shared challenges subsequently becomes an engine for generating the trust necessary to activate social capital.

Koppa observes this phenomenon in the FOCN and other outreach and advocacy programs at RAFI.

> One thing that has proven true in all the work we've done is that farmers listen to farmers more than anyone else. It's a hard job and not something that really makes sense to most people unless you have an immediate attachment to it. So, I think in terms of farmers being able to have a voice in local government, for example, it's very much a question of building power as a group, not as individuals.

As in the other narrative networks profiled in this chapter, land access and tenure have emerged as an important aspect of RAFI's work. RAFI takes a holistic approach to these issues and their overlap with more narrow, conventional definitions of farmland preservation, says Koppa. "The way that a lot of our work is governed is that we think about farming not only as a business or about the individuals who are involved in the farm, but also as historical markers of generational wealth," she said.

> So, I think that the way we approach land tenure is much more from that legal and economic lens, focusing not only on the opportunities that securing tenure to land opens up for people financially but also in terms of other benefits for farm families and the other benefits of attaching family history to a piece of property. Our work is about making sure all those things come together helping our culture move forward into this broader vision of a policy system. It's not just about the land staying in farmland. It's about paying attention to who actually has control of the land and how that land is secured over time.

Like other organizations navigating the dramatic changes in United States agricultural policy instigated by the Trump administration, RAFI is evaluating new ways to serve their constituents. When we spoke in 2025, much of RAFI's financial support from USDA and other federal government sources was in flux, suspended indefinitely, or eliminated entirely. Koppa was convinced, however, that RAFI's broader mission of helping farmers navigate the systematic barriers of the United States agricultural system would live on. "I really see RAFI as very much an anchor organization within this larger ecosystem of American farming," she says. "I think right now people are unclear about who, what, and how to trust because so many things are in flux. So we are trying to make sure that we are one of the places that is still trusted and informed about what is happening."[18]

Narrative Networks and the Politics of Place

"Narrative networks, as we have called them, are simultaneously a story and a grouping of actors," write Lejano et al. "Narratives reveal the complexity of network engagement and the importance of collective and communicative rationality. It the stories people tell us, we find self, other, history, and place all bound up in individual and group narratives."

And place may be the most influential narrative of all, for it bonds us together in a holistic, all-encompassing sense. "Place is a center of meaning constructed by experience," observes Tuan.[19] And though the meaning of those experiences is necessarily subjective, the common experience of place bond us in a powerful ways, no matter our subjective differences.

True, the narrative networks explored in this chapter are not necessarily tied to a specific place, as Kemmis's characterization of the politics of place ascribes, but place can occur at different scales, and the metanarratives of land tenure and land access facing young, BIPOC, and women farmers can be just as relevant in one place as another, even if necessarily flavored by local characteristics and contingencies. This relevance resonates among others in place with similar stories to tell. We listen to stories about our place because it is also a story about our experience, a story about us. This gives the story power and gives us a stake in its outcome. The stories women tell in the Women for the Land Network, or BIPOC farmers tell in the FOCN, are ones of similar challenges and mutual successes. Although they may take place in a virtual meeting space, these challenges and successes occur in a physical place, and their lessons ripple out into local communities. This makes the politics of place more complex in the twenty-first century, more focused on common identity perhaps than common location, but still impactful in local communities and local issues.

Notes

1 Daniel Kemmis, *Community and the Politics of Place* (University of Oklahoma Press, 1990), 135.
2 Raul Lejano, Mrill Ingram, and Helen Ingram, "Environmental Action in the Anthropocene: The Power of Narrative Networks," *Journal of Environmental Policy and Planning* 21, no. 5 (Nov. 2015): 3–4.
3 Raul Lejano, Mrill Ingram, and Helen Ingram, *The Power of Narrative in Environmental Networks* (The MIT Press, 2013), 2.
4 Lejano et al., *Power of Narrative*, 35.
5 Caitlin Joseph, Gabrielle Roesch-McNally, and Ariel Looser, "The State of Gender Equity in U.S. Agriculture," American Farmland Trust (2024), https://farmland.org/blog/inequities-persist-among-american-women-farmers-research-finds.
6 Nicole Gwishiri, "North Carolina Cohort Training on Climate Resiliency Strategies and Local Food Systems," American Farmland Trust, January 5, 2024, https://farmland.org/blog/north-carolina-cohort-training-on-climate-resiliency-strategies-and-local-food-systems.
7 Peggy Petrzelka, Marry Briggs Ott, Ennea Fairchild, and Jennifer Filipiak, "From a Circle of Introductions: Adult Learning and Empowerment of Women Agricultural

Landowners," *Environmental Education Research* 26, no. 2 (2019): 206–218, doi:10.1080/13504622.2019.1632265.

8 Gabriell Roesch-McNalley (Women for the Land Director, American Farmland Trust) in conversation with the author, April 28, 2025.

9 "2021 Transition Recommendations," American Farmland Trust, accessed August 15, 2024, https://farmland.org/transition2021/; and "2022 Census of Agriculture Impacts the Next Generations of Farmers," USDA, accessed August 15, 2025, https://www.ers.usda.gov/topics/farm-economy/land-use-land-value-tenure/farmland-value/https://www.usda.gov/media/blog/2023/02/22/2022-census-agriculture-impacts-next-generations-farmers.

10 "Building a Future with Farmers 2022: National Young Farmer Survey," National Young Farmers Coalition, accessed August 15, 2024, https://www.youngfarmers.org/wp-content/uploads/2022/09/NationalSurveyReport22-1.pdf.

11 "USDA's Increasing Land, Capital, and Market Access Program," National Young Farmers Coalition, accessed August 16, 2025, https://youngfarmers.quorum.us/campaign/protect-LCM/.

12 National Young Farmers Coalition, "Connect with a Young Farmers Chapter," National Young Farmers Coalition, accessed August 17, 2025, https://youngfarmers.org/get-involved/chapters/.

13 "Take Action," Hudson Valley Young Farmers Coalition, accessed August 17, 2025, https://www.hvyoungfarmers.org/take-action-1.

14 "Root and Bloom Fellowship," National Young Farmers Coalition, accessed August 18, 2025, https://youngfarmers.org/get-involved/farmer-fellowships/.

15 David Howard (National Young Farmers Coalition policy director) and Holly Rippon-Butler in conversation with the author, April 18, 2025, and May 7, 2025, respectively.

16 Lejano et al., *Power of Narrative*, 32.

17 Arnaldo Bagnasco, "Trust and Social Capital," in Edwin Amenta, Kate Nash, and Alan Scott, Eds., *The Wiley-Blackwell Companion to Political Sociology* (Blackwell Publishing, 2012), 252–262.

18 Kavita Koppa (Rural Advancement Foundation International executive director) in conversation with the author, May 28, 2025.

19 Y.F. Tuan, "Place: An Experiential Perspective," *Geographical Review* 65, no. 2 (1975), 152.

Bibliography

Bagnasco, Arnaldo. 2012. "Trust and Social Capital." In *The Wiley-Blackwell Companion to Political Sociology*, edited by Edwin Amenta, Kate Nash, and Alan Scott. Blackwell Publishing.

Joseph, Caitlin, Gabrielle Roesch-McNally, and Ariel Looser. 2024. "The State of Gender Equity in U.S. Agriculture." American Farmland Trust. https://farmland.org/blog/inequities-persist-among-american-women-farmers-research-finds.

Kemmis, Daniel. 1990. *Community and the Politics of Place*. University of Oklahoma Press.

Lejano, Raul, Mrill Ingram, and Helen Ingram. 2013. *The Power of Narrative in Environmental Networks*. The MIT Press.

Lejano, Raul, Mrill Ingram, and Helen Ingram. 2015. "Environmental Action in the Anthropocene: The Power of Narrative Networks," *Journal of Environmental Policy and Planning* 21, no. 5. doi:10.1080/19460171.2014.883863.

National Young Farmers Coalition. "Building a Future with Farmers 2022: National Young Farmer Survey." Accessed August 15, 2025. https://www.youngfarmers.org/wp-content/uploads/2022/09/NationalSurveyReport22-1.pdf.

National Young Farmers Coalition. "USDA's Increasing Land, Capital, and Market Access Program." Accessed August 16, 2025. https://youngfarmers.quorum.us/campaign/protect-LCM/.

Petrzelka, Peggy, Marry Briggs Ott, Ennea Fairchild, and Jennifer Filipiak. 2019. "From a Circle of Introductions: Adult Learning and Empowerment of Women Agricultural Landowners," *Environmental Education Research* 26, no. 2. doi:10.1080/13504622.2019.1632265.

Tuan, Yi-Fu. 1975. "Place: An Experiential Perspective." *Geographical Review* 65, no. 2.

16 Crowdlending and Digital Farmland Preservation

It's no revelation our digital world is in many ways undermining American social capital. For better or worse, the "place-ballet" of consistent social interaction that once fueled American community is increasingly moving from the supermarket aisle and church parking lot to the screen and smartphone, undermining the traditional social institutions that once provided society's cornerstone. Kemmis may have endorsed a vision of local, integrated agricultural markets (what he called the "city-state"), starting with the farmers at your local farmer's market, but today's economic landscape is more complicated and digitally integrated than the one he inhabited thirty years ago, and even your local market farmer is probably conducting some share of his or her business online. These digitally oriented New Agrarians are also capitalizing on the Internet's connectivity to empower innovative and sustainable farming operations and expand local agriculture markets through crowdlending, loans made by a group of individuals or institutions recruited through online platforms. Although much of this lending originates from outside the local community, it still supports local farmers and local farm economies, suggesting hybrid models for the politics of place that can function within our new social realities.

These digital financing solutions help address the barriers to financial capital for farmers whose business models operate outside of or adjacent to conventional farming practices. Capital expenses for innovative and sustainable farming practices tend to be high, while financial returns are low, especially in the early years of a new operation. Sustainable farmers often struggle to get loans because traditional lenders tend to favor conventional metrics, lack understanding of regenerative practices, and undervalue ecosystem benefits, while new and beginning farmers lack the experience and financial history necessary to qualify for these loans.[1]

True, the USDA's Farm Service Administration (FSA) offers financing programs that attempt to fill the gaps created by the structural barriers of conventional agricultural financing, but many farmers, both new *and* old, find FSA's bureaucracy almost impossible to navigate, let alone understand. A recent analysis by USDA determined that minority and underserved farmers struggled disproportionately with the requirements of FSA loan programs,[2]

DOI: 10.4324/9781003644545-19

and a National Young Farmers Coalition Survey (NYFC) found that FSA loan programs do not meet the needs of sustainable farmers. The NYFC report recommends the following:

> FSA loan officers should be trained to understand new types of production, including urban and indoor production, that will be important in the face of the climate crisis, and to understand the needs of diversified, organic, and regenerative farming and those receiving price-premiums for local, grass-fed and other premium products.[3]

Internet-based lending and investing networks help mitigate this problem by dispersing the capital pool across like-minded investors and supporters, often at a micro scale. For instance, as part of the research for this book, I invested $250 at an annual return of 6.5 percent to the Steward Regenerative Capital Fund, which provides loans to a variety of regenerative and small-scale agricultural operations across the country. Some of the farms my modest investment supported include a vertically integrated livestock processing cooperative in Montana and a regenerative farm and agritourism education center on San Juan Island in Washington.[4] The $11.12 I will eventually earn on this investment pales in comparison to the emotional satisfaction I receive knowing my money is supporting these types of small-scale agricultural endeavors and helping to preserve the farmland they own and manage.

It may prove that this digitally fueled combination of financial and social capital is the future horizon for farmland preservation in the politics of place, supporting local farmers focused on building social-ecological farming communities through crowdsourced capital. "I think there's just a ton of people out there on the Internet who want to be able to support change in agriculture," says Ryan Anderson, the chief operating officer for the Steward investment platform, profiled below. This is not farmland preservation in the traditional sense, through the lens of protecting the land and its natural and agricultural resources, but farmland preservation in the sense that it harnesses the Internet's inherent narrative networking capacity to empower new and innovative farmers, enhancing their access to both capital and land, and supporting the type of sustainable agriculture that is emblematic of farmland preservation's next generation.

Iroquois Valley Farmland Real Estate Investment Trust

One of the innovators in the field of crowdlending is Iroquois Valley Farmland (IVF), a Real Estate Investment Trust (REIT) that acquires farmland then provides farmer-friendly leases and mortgages to organic, regenerative, and transitioning farmers. REITs are companies that own, operate, or finance income-producing real estate, based on the model of mutual funds. Since Iroquois Valley Farmland's founding in 2007, the investment fund has directed $126 million in organic agriculture investments, most of it invested directly in farmland, as part of its goal to "make organic, regenerative agriculture the norm, not the exception in America."[5]

It all started with one farm in Iroquois Valley, Illinois, and the vision of David Miller, an investment advisor, and his friend, Dr. Stephen Rivard, who combined their interest in organic and sustainable agriculture with a desire to provide people with opportunities to invest in "real" assets beyond the stock market. Miller was leasing a 10-acre farm tract to an organic grower who approached him about acquiring a nearby 160-acre tract so the farmer could expand their operations. "The 160 acres was a little more than I wanted to do," admits Miller, so he approached ten other investors, including Rivard, who eventually became a co-founder of IVF. "Most of the money that came in was from our IRA accounts," explains Miller of IVF's beginnings. "And so we just sort of grew Iroquois Valley Farmland organically. Our first farmer wanted more land and so we bought another farm for him, and we just grew from that base."

Providing expansion capital for existing farmers to acquire farmland has always been at the core of IVF's mission. "We never, ever looked for a particular piece of land," explains Miller. "We have always been reactive to farmers that wanted to expand their business." But at the root of IVF is a deeper mission, to preserve soil through organic agricultural production. "Our vision was to restore the soil back to its natural health," explains Miller. Given their emphasis on soil health, it's no surprise Miller and Rivard are advocates for a healthier American agriculture system. "You know, without our soil, we're finished. We're truly finished as a planet," adds Rivard, a medical doctor who believes many of America's health issues can be directly linked to genetically modified plants, pesticides and herbicides, and other common practices of the commercial agriculture industry. "So, I think that preserving our soil and preserving it healthily is truly the most important thing in America," Rivard says.

To achieve their vision, leases on IVF-owned land require that the lessee transition the land to organic production within five years. "But almost all our farmers were already farming organically and they wanted to do it," explains Miller. "So you know, we were preaching to the choir."

It is through this lens of organic production and soil regeneration that Rivard and Miller see IVF's role in farmland preservationist. Though they are no longer involved in the day-to-day management of IVF, they still play a role in advising the fund's managers, and they said that the company has been discussing selling agricultural easements on their properties for the last several years. "We get that question from our investors all the time," says Miller. "Most that have invested in us and our land would like to see some permanent protections, so that's going to continue to be a 'make-it-better' objective for our company."

For now, however, the company's farmland preservation priorities are centered on clustering investments in local communities to create sustainable organic farming systems. "It's easier to insert a new organic farmer into a community where there's already organic farming being conducted on a large scale," says Rivard. "So that's the next phase of what we're looking at, growing community and clustering investments in organic farmland when the opportunities come up, because what we've learned is that the organic farms that do best are in those clusters." Although they don't mention social capital in discussing this goal, it is implicit in their message—building clusters of sustainable farmers in local farming communities can only enhance the type of relationships and interactions upon which social capital is based.

If IVF is focused on investments in land to support emerging organic farmers and a revolutionary vision for soil health reminiscent of Leopold's "land ethic," their model is essentially a twenty-first century one. Today, IVF's investor pool is no longer limited to Rivard and Miller's circle of friends. The company began offering investments to the public over the Internet in 2019, and as of 2025 had a $127 million portfolio spanning 36,500 acres of diversified farmland (including row crops and pastured livestock) across nineteen states, with a proprietary network of over seventy farmers.[6]

IVF investors now have two investment options: the company's traditional "REIT Equity Shares" and the newer "Rooted in Regeneration Notes" program aimed at supporting disadvantaged farmers. REIT Equity Shares is the company's main investment fund. With a $10,000 minimum investment, and a five-year minimum investment period, the fund gives investors access to a diversified portfolio of loans on organic, regenerative farmland while directly supporting independent farmers with long-term land access, all with an average annualized return of 9.2 percent since its inception.

According to the IVF website, the newer "Rooted in Regeneration Notes" (RNR) investment option "responds to a deep need within the organic, regenerative farming community: supporting social disadvantaged farmers ... who have historically faced steep barriers to land ownership and financing."[7] With a minimum investment of $25,000, RNR investors chose to divert a portion of their return to a shared pool that provides discounted mortgage interest rates to socially disadvantaged farmers whose agricultural operations meet the fund's regenerative farming criteria.

One of the appeals of IVF is that investors can learn about the farms in which they are investing, contributing to the type of social capital created by shared goals and values, even if these investors never meet the farmers they are supporting. A prominent part of IVF's website is its "Farm Stories" section, where the organic and sustainable farms that have received IVF investments are profiled. For example, the RNR program features a profile of Black Wind Farm, a minority-owned organic livestock operation in Albany County, New York. IVF helped owner and Navy veteran Justin Butts finance the acquisition of a 105-acre piece of farmland in 2022. "Traditional mortgage lending wasn't an option for Justin as VA loans don't support commercial farms, and conventional banks often require downpayments beyond reach for new farmers," the IVF website explains.

> On top of his grazing operation, Justin hosts on-farm hog butchery workshops, and organizes a seasonal farm-to-table dinner series ... Justin stewards the land using rotational grazing and organic, regenerative practices, cultivating healthy animals, thriving pastures, and resilient ecoystems.[8]

For Rivard and Miller, this type of investment meets one of their original goals, which was to work "with just average investors who wanted to do something more healthy and impactful for the world and for themselves and for their kids," says Miller, who explains that this is a commitment that works both ways, between investor and farmer.

> At first, we had to go back to the farmers and say, "Guess what, you have investors that trust you." And it worked. Once we had trained our own capital toward this way of thinking, we could boldly tell our farmers that we're going to hold this land for you indefinitely. We actually trained the banks, the bankers, the brokers to believe in this type of investment.

But for Rivard and Miller, this type of hybrid, social/economic capital is only the icing on the cake of their true mission, the preservation of prime, organic soils for the future health benefits of the American public. "When we started, we wanted to offer our investors something real and something healthy, a real asset that you didn't have to sell or trade. That you could just hold onto. And if you farmed that asset organically, it actually grew itself and provided a much more healthy food system for everybody," says Rivard. "Land preservation, you know, and honoring the land and those that came before us, was really critical for us from the start."[9]

Steward

Investors who contribute to IVF's REIT fund have the peace of mind of knowing they are contributing to farmland preservation, with the added comfort of annualized return approaching 10 percent. But they have little say over what specific land they are investing in or where it is located. Like IVF, the investment fund called Steward is adapting popular crowdlending models for investments in regenerative and sustainable agriculture, although at a more egalitarian scale and with more control over investing in specific farms.

Founded in 2017 by crowdfunding pioneer Dan Miller (who previously founded the real estate crowdfunding platform Fundrise), Steward's motto is "Growing a Better Food System," and its mission is promoting environmental and economic stewardship through regenerative agriculture.

Current Steward chief operating officer Ryan Anderson was actually Steward's first investment, and his story crafts a compelling vision of how Internet-based companies like Steward are helping young farmers gain access to land and capital. In 2017, Anderson was a specialty produce farmer in urban Detroit, selling mostly to farmers markets, restaurants, and Community Supported Agriculture (CSA) customers. "I was an urban farmer looking to buy land and expand the business, so I went to a farm credit agency in Michigan," Anderson explains.

> So that was really my first exposure to the sense that ... if you're growing specialty crops, if you're an urban grower, if you're using sustainable and regenerative practices, there's a lot of ways you can quickly fall outside of the credit criteria of those institutions, even if you're running a really successful business.

Denied conventional funding, Anderson soon connected with Miller, who provided him the financing he needed to expand his farming operation. Since then (as of mid-2025), Steward has raised over $66 million in lending, investing in over a hundred regenerative and sustainable agriculture projects across the United States.

They've done this by soliciting short-term loans from groups of individual investors, most requiring a minimum "investment" of only $100. Today, visitors to the Steward website (https://gosteward.com/) are greeted with investment prospectuses for the company's "current campaigns" that feature a brief profile of both the sustainable farming operation the campaign is funding and its farmers. For instance, a recent investment opportunity featured Green Acres Milling, a farmer-owned oat milling operation in Albert Lea, Minnesota. The project is a collective of overs sixty farmers from Minnesota, Iowa, and Wisconsin who are trying to grow the American Midwest's oat industry to rival their successful competition in Canada.[10]

Through crowdlending, Steward successfully raised $2.389 million for the company, which needed the funding to match a USDA Regional Food System Infrastructure (RFSI) grant so it could purchase specialized equipment to process sustainably grown organic oats grown from locally adapted oat seed. As with its other investment opportunities, Steward offered a modest return (here a 10.5 percent annual percentage rate on a nine-month term, which appears more or less typical based on an anecdotal review). But the return on investment is only part of the appeal, because investors also get a chance to learn about the farmers and sustainable farming practices their investment is supporting. For instance, the Green Acre Milling prospectus introduces its investors to Anne and Landon Plagge, the managers of the Green Acres Milling cooperative, and describes the regenerative farming practices they use.[11] "So on our website, you can go and read about the different projects seeking funds, and you can do your own underwriting on the environmental and economic impact for those communities," Anderson says about the appeal of the website's farmer profiles. "And I think there's just a ton of people out there who want to be able to push that kind of change forward, and if they're going to support change in agriculture, they like to know exactly what it is they're supporting."

Anderson describes Steward's business model as a participation lending model with two sides. "On the one side are the borrowers running these amazing agricultural enterprises," he explains. "On the other side we have a network of a few thousand private lenders that are actually amassing the capital. And Steward is in the middle, combining all those participations into one single loan that it manages." According to Anderson, functioning as a private lender gives Steward more flexibility to structure its loans in a way that makes sense for their borrowers but also focus their investments. "One of the things that is unique to us is what we call our 'stewardship standard' where we are looking at the environmental and economic social practices of the farming operation," he explains. "We're really trying to assess whether this operation is improving the way American agriculture operates. Do they care about soil health? Do they care about biodiversity? Are they paying their employees a living wage?"

Farmers and agricultural businesses seeking investments can apply through the website, but Anderson says the majority of their borrowers are referrals from their existing loan clients. "Once we've worked with a farmer and it goes well, they appreciate the perspective we have and are recommending us to people they know in similar situations." Although the company does evaluate loan opportunities based on a scorecard of financial criteria, Anderson says the company's evaluation process also includes qualitative criteria regarding the businesses' commitment to regenerative farming practices and improving local farm economies.

Anderson points to two main ways that Steward is contributing to the farmland preservation movement. One is through facilitating the transition of farms and farm businesses between generations. For instance, Anderson highlights a recent investment opportunity it offered for StarWalker, an organic farming and beef operation in northern California's Scott Valley run by a couple named Jason and Christina Walker.[12] Part of the investment package was dedicated toward helping the Walkers buy land they are currently farming from Jason's parents, who needed to sell part of their land to generate capital for their retirement plans. "That need to generate capital through land assets creates all of these generational succession challenges for families that want to be able to keep that land in the family and in agriculture," says Anderson. "So we're working pretty hard to be flexible and strategic in how we structure loans in order to support farm families who are going through that transition."

The other side of Steward's farmland preservation work is helping farmers keep their businesses sustainable and profitable by providing the financial capital such efforts typically require, Anderson says. Keeping farms economically viable inspires future generation of farmland owners to stay on the farm, enabling them to see a financial future for themselves in agriculture—the heart of the "agricultural viability" issue which has been discussed throughout this book.

> So much of succession planning is just about stabilizing the finances of these farming operations and recognizing what new structures small farmers need to survive moving forward. And so we do a lot of work with small farms where they are adding on-farm processing. And it's as much about having something that makes sense for the future generation as it does for the ones operating it now.[13]

But Anderson is perhaps prouder of the broader social and economic benefits the Steward investment platform offers to the rural communities it serves with its lending platforms. "As we operate with our lending and supporting the development of these supply chains, we want to find ways of driving economic value back to the farmers, and then through them into those rural communities where they are located," he says. "In order to keep agriculture in America, to keep small farms, to keep land in farming with those small farms, we need new business models. And Steward wants to be a part of that effort."

The argument that Internet-based crowdlending and investment platforms like Steward and IVF can emerge as important components of farmland preservation in the politics of place may sound grandiose to some, but thinking about these platforms through some of the theoretical concepts explored elsewhere in this book perhaps opens new pathways of perception about farmland preservation. Like bridging organizations, these Internet investing platforms connect hierarchies and nodes of the farmland preservation community through digital networking and access to financial capital, an increasingly important resource for the small and medium-scale farmers who typically chose to preserve their farmland. Like narrative networks, they use stories of famers, farmland, and community to attract investors and support new and expanding rural enterprises, thus supporting the rural communities where these enterprises operate. As Anderson says when we close our interview,

> I think the conversation around the impact that agriculture is having on our environment and the impact it's having on our rural communities is only going to grow and continue. There's just so much opportunity both in the United States and globally to change how we're thinking about the role of agriculture in society, and Steward wants to be a part of that conversation.

Notes

1 Moises Barbosa Jr., Elian Pinheiro, Carla Sukulski, Diego Alexis Huarachi, and Antonio Carlos de Francisco, "How to Identify Barriers to the Adoption of Sustainable Agriculture?" *Sustainability* 14, no. 20 (2022): 13277.

2 "Shaping Change at USDA: Final Report," USDA Equity Commission, February 2024, https://www.farmaid.org/wp-content/uploads/2025/02/usda_equity_commission_final_report-2024.pdf.

3 "Building a Future with Farmers 2022: Results and Recommendations from the National Young Farmer Survey," National Young Farmers Coalition, 2022, https://youngfarmers.org/wp-content/uploads/2025/07/National-Survey-Web-Update_11.15.22-1.pdf.
4 "Steward Regenerative Capital Fund," Steward, accessed December 12, 2025, https://gosteward.com/projects/steward-regenerative-capital.
5 Iroquois Valley Farmland REIT, accessed July 22, 2025, https://iroquoisvalley.com/.
6 "About Us," Iroquois Valley Farmland REIT, accessed December 15, 2025, https://iroquoisvalley.com/about/.
7 "Rooted in Regeneration Notes," Iroquois Valley REIT, accessed December 15, 2025, https://iroquoisvalley.com/invest/rooted-in-regeneration-notes/.
8 "Farm Profile: Black Wind Farm," Iroquois Valley REIT, accessed December 15, 2025, https://iroquoisvalley.com/2025/08/07/farm-profile-black-wind-farm/.
9 David Miller and Stephen Rivard (Iroquois Valley REIT) in conversation with the author, May 1, 2025.
10 Evan Berg, "Green Acre Milling to build new oats processing facility in Albert Lea," KIMT3 News, November 15, 2024, https://www.kimt.com/news/green-acres-milling-to-build-new-oats-processing-facility-in-albert-lea/article_269f88a2-a3ac-11ef-892a-37fdfa4d1a9f.html.
11 "Equipment Purchase for Farmer-Owned Oat Mill," Steward, accessed July 25, 2025, https://gosteward.com/projects/green-acres-milling/equipment-purchase-for-farmer-owned-oat-mill.
12 "Working Capital to Scale Direct to Consumer Sales," Steward, accessed July 25, 2025, https://gosteward.com/projects/starwalker/working-capital-to-scale-direct-to-consumer-sales.
13 Ryan Anderson (Steward) in conversation with the author, May 22, 2025.

Bibliography

Barbosa, MoisesJr., Elian Pinheiro, Carla Sukulski, Diego Alexis Huarachi, and Antonio Carlos de Francisco. 2022. "How to Identify Barriers to the Adoption of Sustainable Agriculture?" *Sustainability* 14, no. 20 (2022): 13277. doi:10.3390/su142013277.

Iroquois Valley Farmland REIT. n.d. https://iroquoisvalley.com/.

National Young Farmers Coalition. 2022. "Building a Future With Farmers 2022: Results and Recommendations from the National Young Farmer Survey." https://youngfarmers.org/wp-content/uploads/2025/07/National-Survey-Web-Update_11.15.22-1.pdf.

USDA Equity Commission. February. 2024. "Shaping Change at USDA: Final Report." https://www.farmaid.org/wp-content/uploads/2025/02/usda_equity_commission_final_report-2024.pdf.

Steward. n.d. https://gosteward.com/.

17 International Farmland Preservation

Although this book focuses on its American policies and practices, farmland preservation is a global issue. Climate change, food security, and the preservation of traditional lifestyles and landscapes are all reasons countries around the globe attempt to limit or prevent the conversion of farmland and the degradation of prime soils, forests, and other natural resources associated with agriculture.

Driving much global emphasis on farmland preservation is new awareness about the impacts of climate change and the role farm and working forest lands can play in its mitigation. A 2024 report by the American Farmland Trust notes that permanently protected farmland can reduce greenhouse gas emissions and contribute to climate resilience, among other environmental benefits.[1] In particular, regenerative farming practices—such as no-till farming, cover cropping, rotational grazing, and agroforestry—can help sequester carbon, improve soil quality, and protect drinking water supplies. "Simply put, recent data from farming systems and pasture trials around the globe show that we could sequester more than 100 percent of current annual CO_2 emissions with a switch to widely available and inexpensive organic management practices," states another recent report by the Rodale Institute.[2]

The issue of food security is perhaps even more dire than climate change, at least in the short term, for countries experiencing rapid population growth or saddled with inhospitable agricultural climates. In the last hundred years, global population has increased over fourfold, leaving the food supplies of densely populated countries like Syria, Haiti, and Sudan in crisis, and forcing heavily populated nations like Egypt and Bangladesh to institute robust food system planning policies.[3] Water scarcity, urban sprawl, and threats to indigenous tribes and communities are other threats driving attention to farmland preservation across the globe.

However, just as farmland preservation in each state of the United States is not quite the same, driven by state and local policies, farmland preservation across the globe is driven by the *esprit* and *esprit de lois* of each individual country, and the climate, religion, government systems, and cultures that define them. Indeed, the study of farmland preservation across the globe reveals the "politics of place" in full effect, at least on a national scale.

DOI: 10.4324/9781003644545-20

In Great Britain and other northern European countries such the Netherlands, strong central planning and national *moeurs* favoring urban containment and rural landscapes have led to robust farmland preservation policies. China also relies on strong central planning for farmland preservation, though their system focuses on maintaining the balance between food system demand and supply through comprehensive production quotas, reflecting the *moeurs* of a country that has long struggled with food security for its massive and politically isolated population. In Canada, like America, farmland preservation policy is more region- or province-specific, fashioned by urban growth and sprawl in some places and cultural and political sentiment in others.

While a comprehensive review of international farmland preservation policies and practices is worthy of its own book-length treatment, this chapter provides a selected review of such policies from around the globe.

England and the Netherlands

England and the Netherland both have high population densities, each ranking within Europe's top ten.[4] Both countries embrace comprehensive regional and local planning initiatives, directed by national policy and planning oversight. Both share strong national *moeurs* on the importance of maintaining a balanced relationship between urban areas and the surrounding countryside. And in comparison to the United States, both share robust farmland preservation policies at national, regional, and local levels.[5]

In England and Wales, regional and local planning in rural and agricultural areas is administered by the Department for Environment, Food and Rural Affairs (DEFRA), who's national "Agricultural Land Classification" (ALC) system governs farmland preservation and conversion. Farmland classification in the ALC is based on climate, site, and soil characteristics, and the important interactions between them. Under DEFRA's overall policy direction, approval for nonagricultural development on lands classified as Grades 1 (Excellent Quality Agricultural Land), 2 (Very Good Quality Agricultural Land), and Subgrade 3a (Good Quality Agricultural Land) is required by Natural England, England's primary national agency for conservation and environmental preservation. Changes to uncultivated or semi-improved land, or projects that restructure rural land, are governed by the Environmental Impact Assessment Regulations of 2006, while forestry related conversions are covered by the Environmental Impact Assessment Regulations of 1999.[6]

In 2018, England published "A Green Future: Our 25 Year Plan to Improve the Environment," which provides national direction for a variety of environmental initiatives, including agricultural sustainability. According to its executive summary, the plan calls for "an approach to agriculture, forestry, land use, and fishing that puts the environment first" and attempts to seize a "once-in-a-lifetime chance to reform our agriculture and fisheries management, how we restore nature, and how we care for our land, our rivers, and our seas." Although the plan does not specifically address farmland preservation practices, it acknowledges the role farmland plays in tourism and recreation, air quality, and habitat preservation, among other environmental benefits it hopes to facilitate.[7]

England's farmland preservation policy is also guided by its National Planning and Policy Framework (NPPF), which directs local planning authorities (LPAs) to consider farmland's role in protecting and enhancing rural landscapes, biodiversity, and soils, and preventing "land instability" among other environmental objectives when considering regional and site-specific development proposals.[8]

England's strong central planning authority and oversight for farmland preservation is aided by its legal framework regarding private property rights and permissions. Unlike America, English landowners do not possess broad "inalienable" rights to develop or subdivide their property, nor the *esprit de lois* that institutionalizes such rights. This difference in legal traditions helps explain the relative insignificance of conservation easements as a tool for farmland conservation in England. They are, one could argue, inconsequential in a country that both has so many other farmland preservation policies and largely lacks America's property rights zealotry. Although England's Environment Act of 2021 enabled "conservation covenants"—private landowner agreements that mimic American conservation easements—their use has only been in effect since 2022 and are therefore still relatively small and unstudied.[9]

What is prevalent is a nationwide *esprit* valuing rural landscapes and the containment of urban sprawl. The inspiration for this *esprit* sprung from the ills of the industrial revolution—disease, overcrowding and poverty, environmental pollution—coupled with the Victorian concept of "Garden Cities," a vision for self-sufficient, self-contained towns using planned green space to alleviate urban pollution and congestion and enhance social well-being. This vision is epitomized by the famous "City Cluster" diagram of Ebenezer Howard, an English urban planner who popularized the Garden City movement. This diagram depicted a radial distribution of small cities around a larger central city, each separated by a greenbelt of rural, agricultural, and natural spaces.[10]

Although Howard's Garden City concept was always a utopian ideal, its influence resulted in a series of experimental suburbs surrounding London, including Bedford Park, Hampstead, and Ealing Garden, combining garden and open space with cooperative housing models. Howard's ideas were instrumental in forming the London County Council, a forum for the strategic planning of London's urban fringe, and eventually the launching of England's Greenbelt movement. In 1938, this social movement resulted in passage of the Green Belt (London and Home Counties) Act, which severely limited construction and development in greenbelt-designated areas on London's outskirts, granted statutory oversight over these areas to planning authorities, and allowed local governments and parish councils to acquire land for greenbelt development. "Local government was remarkably creative in securing land for Green Belts ... There seems little doubt that this was a broadly popular movement driven by creative government for the public good," writes author Peter Bishop. "The Green Belt is one of the few planning measures in the UK that has entered the public consciousness ... 'Green Belt' is a powerful brand name that garners broad support."[11]

Today, not everyone is a fan of England's Green Belt programs, especially as the country's affordable housing crisis deepens. Recently, British prime minister Keir Starmer advocated for loosening Green Belt protections to permit new homes on lower-quality Green Belt land such as unused playing fields. Starmer's opponents, meanwhile, eagerly defended Green Belt protections, calling instead for new home construction on the brownfield sites of previous industrial and commercial development.[12] These debates reflect the view of researcher Rachelle Alterman, who in her review of British farmland preservation policy notes, "Green belts have assumed enormous political and emotional significance."[13] And though she made this observation almost twenty years ago, the adoption of the "Green Future" plan in 2018 suggests the Green Belt ethic is still strong in English consciousness.

Like England, the Netherlands primarily achieves farmland preservation through a patchwork of environmental policy and planning provisions designed to control rural sprawl, promote biodiversity, and maintain the country's rural and natural character. Examples of the Netherlands hybrid approach to farmland preservation policy include the National Ecological Network, designed to enhance connections of natural areas and surrounding farmland,[14] and the 2024 Environment and Planning Act, which integrates dozens of complex land use, planning, and environmental laws into a single, comprehensive framework to protect water supplies and support biodiversity goals, with farmland preservation as a substantial secondary benefit.[15]

The Netherlands is one of the most densely populated countries in Europe, with a population of 18 million people in a country roughly the size of West Virginia. As of 2021, the country had the second-highest population density in Europe, at 521 people per square kilometer.[16] The Netherlands also has one of the most industrialized and intense agricultural systems in Europe, with about half of its total land area in agricultural production as of 2020, ranking fourth in the European Union in that statistic behind Ireland, Denmark, and Spain.[17]

Despite its intense urbanization pressure, the Netherlands is considered a success story in European farmland preservation, primarily due to its strong planning policy framework that concentrates urban growth in regional centers. This framework is supported by close coordination between the Ministry of Agriculture, Fisheries, Food Security, and Nature (LVNN) and the provincial and local governments responsible for implementing national land-use policies.

Open space planning in the Netherlands is supported by a variety of incentive programs for agricultural landowners. In 2016, the country introduced a new collaborative agri-environmental framework to coordinate distribution of agricultural incentive and subsidy programs through agricultural collectives, which are groups of farmers and other rural landowners organized as publicly certified conservation organizations. As of 2023, there were forty agricultural collectives in the Netherlands, organized around regions and covering the entire Dutch countryside. The mandate for these collectives not only includes the distribution of financial incentives for sustainable environmental management of farmland and natural areas, but also inspecting and, if necessary, sanctioning individual landowners within its regional boundaries for program compliance.[18]

The Netherlands also provides landowners with tax incentives, similar to the "use-value" or deferred assessment programs in America. Under the Dutch Nature Conservation Act, passed in 2017, land or estates that are at least 5 hectares in size and have over 30 percent forested or natural areas (or smaller estates with historical significance for an adjacent one) may qualify for a suite of tax benefits relating to income tax, inheritance tax, property tax, and other types of tax liabilities.[19] Although these benefits are specifically tied to natural area preservation, farmland preservation is often an auxiliary beneficiary, as they allow farmland owners to attain significant tax benefits for maintaining natural areas adjacent to their farmland, thus supporting a holistic, ecosystem approach to farmland preservation.

Canada and Australia

Given Canada and Australia's similarity with the United States in population, cultural history, and the legacy of English common and real estate law, one might presume their farmland preservation policies are also similar to the United States. Indeed, tools similar to the American "conservation easement" are prominent in both countries, and their use is growing.

In Australia, these private conservation agreements are called "Privately Protected Areas," or PPAs, which are restricted and stewarded according to a conservation covenant—essentially a legal agreement similar to a conservation easement. As of 2024, there were 6,770 PPAs in Australia, protecting over 10.6 million hectares of various land types, including agriculture and rangelands.[20]

Most Australian conservation covenants are voluntary, with the landowner agreeing to a suite of restrictions with the local or regional agency, and the agency assuming the legal defense and stewardship role. Provisions for stewardship and enforcement of Australian PPAs vary widely, with most conservation covenants overseen by a dedicated agency within a state government or a state-funded and controlled statutory body. Although there is a burgeoning private land trust model in Australia, most of these private conservation groups are focused on wildlife and natural habitat conservation, not farmland preservation.[21]

Recent totals for privately protected land in Canada could not be found, but every Canadian jurisdiction or territory authorizes conservation easements, or similar, private land protection agreements, although private land conservation policy differs wildly depending on the regional or territorial governance. Ten Canadian jurisdictions (territories or provinces) have explicit legislation that allows for use of conservation easements, while three have historic resource laws that allow for permanent conservation of private land. Like America, most conservation easements or private land conservation agreements are held by land trusts, such as Ducks Unlimited or The Nature Conservancy (although local and regional land trusts are growing), while some are held by local and regional governments and other types of environmental nongovernment organizations (NGOs).[22]

Despite their adoption of American-style conservation easements, Canada and Australia differentiate themselves from the United States through the stronger role regional and provincial planning plays in farmland preservation there. Population density also strongly influences these regional policies. Canada and Australia's population density is each roughly four people per square kilometer, compared to the United States density of thirty-six people per square kilometer. In Australia, population density is centered in a small number of large metropolitan areas in its coastal region. In Canada, population density is similarly centered in a few metropolitan areas, while vast swaths of isolated rural areas characterize most parts of the country. This leads to broad differences in farmland preservation policy across Canada's provinces and territories.

The Canadian farmland preservation movement is centered primarily in the more highly urbanized Ontario, Quebec, and British Columbia provinces, which also happen to be the home of much of the country's best farmland. In Ontario, where the rate of farmland conversion is the most alarming, farmland preservation policy is directed by a series of planning acts requiring compatibility of local development plans with provincial priorities. The most important planning policy is the "Official Plan," which guides development and preservation decisions and must be approved by the Ministry of Municipal Affairs and Housing. As part of the Official Plan process, municipalities and regions are obligated to identify and protect prime agricultural areas and may not allow urban expansion into these areas unless it can be demonstrated that there is no reasonable alternative for avoiding them.

Livestock and grazing remain important agricultural sectors in Australia's rural, remote regions. But like Canada, much of Australia's prime soils and viable farmland are also located in the same coastal regions as its densely populated urban areas. And these coastal farming regions are increasingly threatened by issues such as water scarcity, major mining expansion, the need for affordable housing, and urban sprawl.

Although interest in preserving Australia's prime farmland is on the rise, the authors Trevor Budge and Andrew Butt characterize this interest as reactive, centered primarily around these urban areas, with only mild success so far. "The importance of peri-urban agriculture for a healthy city and the advantages of retaining agriculture are being increasingly recognized in rhetoric, but not in planning strategies," they complain. "These objectives have generally been driven by concerns over the implications of such a pattern for transport access and costs and the provision of infrastructure, facilities, and services, rather than deep concern about encroaching on productive agricultural land."[23]

Australian farmland preservation efforts are complicated by the country's growing water crisis, with the debate driven by difficult decisions regarding water allocations for irrigation and other agricultural production versus those for environmental and human consumption needs. But farmland preservation remains an important issue to the Australian people, and Australian governments have responded with a matrix of planning policies designed to facilitate more balanced, sustainable land use.

An Australian "National Food Plan" with a goal of fostering "a sustainable, globally competitive, resilient food supply" was adopted in the 2010s but never successfully implemented.[24] The initiative was revived in 2025, however, when the Australian government proposed its "Feeding Australia" national food security strategy. However, as of the publication of this book, that plan is still in the development and consultation stage.[25]

With little comprehensive national guidance, farmland preservation has fallen either to the PPAs described above or planning policies crafted by the individual Australian states, with day-to-day administration delegated to local governments. For instance, the state of Queensland—home to Brisbane, the country's third-most populous city—adopted a "Strategic Land Cropping Act" in 2013 that attempts to protect land with prime soils and manage development impacts in those areas. However, the policy only requires local governments to "seriously address" the preservation of these areas in development proposals, with few mandatory protections.

Certainly, not all of Australia's farmland preservation policy is ineffectual. In 2002, the Melbourne metropolitan region initiated the "Melbourne 2030 Plan," an extension of regional planning strategies developed during the late 1960s. Now Australia's most populated city, Melbourne is also surrounded by some of the country's most productive and threatened farmland. The "2030 Plan" attempted to confine future urban growth to three major corridors on Melbourne's metropolitan fringe, essentially establishing a greenbelt around the city. "Even without a specific stated intention to preserve farmland, implementation of the Melbourne strategy has potentially delivered the most significant agricultural protection measure anywhere in Australia," write Budge and Butt. "The means of enforcement through a legislated urban growth boundary are generally considered to be the strictest provisions in metropolitan planning around the world."[26] However, not all evaluations of the Melbourne greenbelt share this sanguine outlook, and a recent case study published in 2024 suggests the "strict provisions" of Melbourne's greenbelt program have been slowly eroded by compromise and an ever-expanding list of exceptions, along with the perhaps inevitable fragmentation of large, undeveloped tracts.[27]

Like Melbourne's initiative, and similar American efforts, the Canadian provinces of Quebec and Ontario attempt to control urban sprawl and farmland conversion through mandated, greenbelt-like planning protocols around major cities such as Quebec City, Montreal, and Ottawa. The Ontario policy is the more decentralized of the two, requiring the compatibility of local plans with provincial priorities in farmland protection programs. All local and regional municipalities in Ontario must adopt official plans addressing provincial farmland preservation priorities. Similarly, all plans must be approved by the Provincial Minister of Municipal Affairs and Housing, while implementation policies such as zoning bylaws and subdivision ordinances must conform to higher-level planning documents and provincial priorities.

Quebec, on the other hand, has established a special provincial commission to designate urban and protected zones for each municipality in the province and regulate subdivision. The commission alone has authority to permit changes from its own designations. Alterman suggests that Ontario's autocratic approach to farmland preservation is motivated by "nationalistic French-Canadian views of the possible need for self-sufficiency in the event of succession," along with the still-potent political influence of its farmers and rural groups.[28]

Despite their aggressive goals, however, the Ontario and Quebec policies have struggled to create meaningful farmland preservation outcomes against the relentless pressures of urbanization. Yet both can be deemed successes, if conditional ones. A 2022 study analyzing farmland loss from 2000 to 2017 in Central Ontario, a region containing some of Canada's most productive soils but also some of its most intense urban pressure, found that its greenbelt policies combining both provincial and local planning frameworks had significantly reduced the rates of farmland loss since 2005, while other planning policies contributed to enhanced municipal control over agricultural land conversion.[29] Similarly, a 2021 study determined that Quebec's policy framework remains the strongest farmland preservation initiative among all Canadian provinces.[30]

China

Perhaps no country has been as serious or systematic about farmland preservation as China. With the second-largest population in the world (behind India) and relatively small amount of prime farmland, China's central government has implemented strict farmland preservation requirements since 1990s in response to the country's rapid urban expansion.

Preceding China's farmland preservation policies of the 1990s was an era of farmland enhancement and engineering designed to increase agricultural production for the country's expanding population. These measures included building terraces, implementing water conservation protocols, and enhancing soils, resulting in a 5.3 percent annual average increase in grain production from 1963 to 1978.

However, ongoing industrialization and urban expansion still reduced China's overall farmland area at an average annual rate of 0.39 percent from 1979 to 1989. In response, China's national government mandated farmland preservation land management systems called Basic Farmland Protection Systems (BFPS) for all parts of the country in 1994. In 1998, China's Land Management Law defined "basic farmland" as "farmland that must not be occupied by non-agricultural land use in accordance with the demands of the population and the requirements of socioeconomic development for agricultural products," further codifying BFPS in China's land management systems.[31]

According to BFPS regulations, land classified as basic farmland can only be used for growing major food and feed grains. Basic farmland criteria are based primarily on high soil productivity, including the quality and stability of the land and suitability for cultivation. Criteria also consider socio-economic factors such as supporting long-term national plans and meeting demand for agricultural product. As of 2017, approximately 76 percent of China's total farmland was designated basic farmland.

Although the strict provisions and enforcement of China's basic farmland policy may seem like a farmland preservationist's (fever) dream, the program has endured criticism in recent years. The policy subjects farm households to heavy punishments for changing the use of their land or the type and variety of crops grown, limiting their ability to adapt to changing markets. However, regional differences exist. Since the country's major grain-producing regions are located mainly in eastern and central China, western Chinese farmers have enjoyed greater freedom to choose crop types and change their land use, facilitating more entrepreneurial practices. Still, the BFPS policy results in rigid control over land use throughout China, with strict government oversight and enforcement.[32]

In 2025, the Chinese government announced a target of adding 1.35 billion mu (approximately 222 million acres) of new basic farmland to meet estimated long-term agricultural production needs through its new "high standard farmland" initiative. The BFPS system already included measures for enhancing substandard farmland through soil improvements and water conservation, among other practices. Still, given the intense production demands placed on existing and converted basic farmland, maintaining farmland quality is a challenge throughout the country. The "high standard" initiative aims to overhaul substandard land by leveling fields, enhancing drainage and irrigation, consolidating fragmented plots, and improving access for farm machinery and other evolving agricultural technology.

Critics argue this massive expansion can only be achieved through the damaging conversion of wetlands, grasslands, and other environmentally fragile areas. Supervision and enforcement of BFPS policies remains difficult over such a vast territory. Meanwhile, the harsh realities of China's growth continue, with one study determining that construction related to housing and urbanization was the main factor causing farmland loss from 2010 to

2017. In response to this market imbalance, the Chinese national government experimentally decentralized the approval authority for converting permanent basic farmland to provincial-level administrative regions in 2020. Others complain that the BFPS is not keeping up with changing market needs and demand. China's demand for grain is declining, with a grain surplus predicted for 2025, while Chinese consumers demand for other agricultural products such as meat, eggs, milk, and fruit is on the rise.[33]

China is a developing country with a large population and proportionately little farmland. The conflicts between people and land preservation are intense, and unlikely to improve significantly, despite the country's comprehensive efforts at farmland preservation. As the Chinese government navigates these challenges, its critics fear that other environmental protections will fall by the wayside. Indeed, the environmental legacy of the People's Republic of China is already a sad one. Some suggest these challenges can be met through an even more sophisticated planning regime, a more efficient and integrated review process, and more stringent monitoring of plan implementation.[34] Still, the example of China shows that systematic farmland preservation is possible on a national scale, even if such systemization seems unlikely in countries with more liberal legal rights and democratic principles.

Israel

With a 2025 population density of 440 people per square kilometer,[35] Israel's is both the highest population density and smallest national land area covered in this review. It also has the highest birth rate among industrialized countries, reported at 2.9 children per woman in 2024.[36]

These demographics are almost unimaginable in the United States, yet like America, farmland plays a unique role in the *moeurs* and symbolic capital of Israeli culture, dating back to the country's founding as a refugee state and the influence of the Zionist movement. This cultural heritage led to robust farmland preservation policy in the early mid- to late twentieth century, overseen by the country's national government. Yet like other industrialized countries, urbanization and population growth have significantly undermined this legacy of farmland preservation in Israel, despite a growing movement to preserve the country's rural character in more recent years.

To understand Israeli farmland preservation is to become familiar with its Zionist ideology and the "Land Narrative" embedded in it. The Zionist movement was aimed at creating a Jewish national state and homeland in Israel; underlying it were socialist values idealizing a cooperative society based on farming and self-sufficiency. This ethos led to agricultural land preservation promoting cooperative and rural settlement as one of Israel's foundational principles. As a result, most Israeli land (93 percent) is nationally owned, making it unique among Western countries. This land is owned either directly by the state or by quasi-governmental bodies authorized to develop and oversee land, such as the Jewish National Fund (JNF) and the Israel Lands Administration (ILA), whose policy is directed by the Israel Lands Council (ILC).[37] Israel's 1965 Planning and Building Law mandated the protection of agricultural lands as a statutory objective of local, district, and national plans and also established an eleven-member Committee for the Protection of Agricultural Land (CPAL), which was authorized with final approval over all requests for agricultural land conversions (a body scholar Rachelle Alterman calls "the ultimate farmland conserver's dream").[38]

The scholar Ravit Hananel refers to this *moeur* of rural farmland preservation as Israel's "Land Narrative," institutionalizing the preference of Israel's early government for "the cooperative and rural sector over the urban sector. According to the narrative, rural-agricultural settlements and settlers are preferable from the national standpoint to urban settlements and those who live with them."[39]

In 1965, the ILC adopted the "Land Policy of Israel," which created basic lease categories for Israeli land—one for urban areas and the other rural-agricultural land. The "Land Narrative" was embodied in the extreme price controls for rural-agricultural leases (initially based on 2 percent of the homestead's net revenue). In return, rural lessees were obligated to permanently reside on the land and use it for agriculture. If these two conditions were not met, the ILA had the authority to void the lease and seize the land. Moreover, the rural lessee could not divide the land among descendants but was required to convey family land rights to one "continuing" heir in order to preserve the land's agricultural viability.

During the 1990s, however, a variety of factors led to a loosening of Israel's strict agricultural land protections and a transition toward more private property rights for rural landowners. One factor was a massive influx of Jewish immigrants from the newly independent states of the Soviet Union. Another was a debt and profitability crisis suffered by many in Israel's agricultural sector, centered in its cooperative kibbutz and moshav communities. Critics argued the crisis was exacerbated by the CPAL's overzealous farmland protections, showing it was out of touch with the country's changing priorities and *moeurs*.

Coupled with ongoing population growth, housing and development pressures, and a general trend towards privatization in Israeli culture, this crisis led to a series of reforms that gradually undermined Israel's robust farmland preservation policies. In an effort to help struggling farmers, the ILC made a series of decisions allowing farming settlements to convert agricultural land for industrial, commercial, and residential purposes. These reforms culminated in a 1996 revision to the "Planning Law" that reduced the CPAL's influence over farmland conversion requests.

Another significant reform was 2009 legislation that opened a pathway to the privatization of urban lands. Although the national ownership structures of rural-agricultural land remained largely unchanged, scholar Ravit Hananel complains that the 2009 reforms reflect a broader neoliberal trend that ultimately undermines protections for the Israel's farmland and agricultural system.[40] Indeed, farmland conversion in Israel has accelerated since 2000, with cumulative farmland loss since then estimated at 100,000 to 125,000 acres, while the Ministry of Agriculture and Development estimates farmland loss in the 2020s at approximately 3,700 to 5,000 acres per year.[41] Meanwhile, thanks to technology and mechanization, the Israeli agricultural sector is trending toward larger farms focusing on a smaller number of crops with a smaller number of farm workers. Globalization and imported agricultural products have also decreased agriculture's overall contribution to the Israeli GDP over the last few decades, declining from 2.4 percent of GDP in 2000 to 1.3–1.4 percent in recent years.[42]

Given the realities of Israel's population growth and housing needs, farmland conversion in Israel's rural areas is perhaps inevitable. But compared to America, the country still maintains significant control over farmland conversion, and the maintenance of rural areas and rural character remain a strong preference in Israeli culture. In part, this is due to the unique social structures of the Israeli agricultural communes, based on the Zionist "land narrative" of a self-sustaining Jewish people. And thanks in part to Covid, new generations of middle and upper-class Israelis are seeking rural lifestyles, a development scholar Gabriel Schwake calls "neo-rurality," which he characterizes as the trend of Israelis moving away from urban centers to seek better living standards while searching for a more tranquil way of life.[43]

As these urban refugees seek rural housing, this "neo-rurality" has led to new development pressures on farmland, along with new interest in smart-growth policies that attempt to balance growth with preservation of rural character. As a result, Israeli planning policies have shifted from a focus on agricultural and ideological land use to those emphasizing protection of open spaces for cultural heritage and environmental benefits. Even as agriculture's role in Israeli culture declines, modern Israeli planners seek to utilize small-scale farmyards and agricultural remnants in urbanizing areas as open space infrastructure for the country's increasingly urbanizing countryside.[44]

Brazil

Although this review has focused on highly developed countries such as England, the Netherlands, and Israel, farmland preservation is also taking place in developing countries such as Brazil. However, Brazilian farmland preservation policies are typically framed within larger environmental concerns and narratives, rather than the "New Agrarian" narratives of American farmland preservation policy.

For example, the Brazilian Forest Code, revised in 2012, requires rural landowners to maintain a percentage of their forest as set-asides and riparian buffers to balance industrialized agriculture uses with ecological conservation. Similarly, the Agricultural Zoning of Climate Risk (ZARC) policy attempts to mandate sustainable crop selection based on soil, rainfall, and temperature assessments to preserve prime farming areas. Currently, Brazil does not allow American-style conservation easements, but it is attempting to motivate conservation management and environmental protections through financial incentive programs. Recently implemented Payment for Environmental Services (PES) programs compensate agricultural landowners for adopting conservation practices.[45]

Since the 1980s, Brazil has also attempted to implement policies for agrarian reform and land regularization aimed at clarifying and formalizing land tenure for small farmers. For instance, Brazil's National Institute of Colonization and Agrarian Reform (INCRA), the country's main agency for land reform for small-scale farmers, has overseen the settlement of over 900,000 families on more than 80 million hectares of redistributed land since the mid-2010s. Similarly, land regularization policies such as the Terra Legal Program attempt to formalize land tenure in the Amazon and other frontier regions.[46]

Sadly, these policies have been characterized by a legacy of lax enforcement, corruption, and amnesty provisions that can encourage violations. Land distribution in Brazil remains unequal, concentrated in the hands of a few wealthy landowners who have little incentive toward conservation practices, and agrarian reforms have slowed in recent years, threatening tenure and land access for poorer, small-scale farmers. The limitations of Brazilian farmland preservation efforts show that no policy can be successful without the willpower to confront powerful economic and political interests in less developed countries.

Conclusion

Writing almost three hundred years ago, Montesquieu believed that the circumstances of a country or place—its laws, commerce, culture, geography, and physical environment—best determined the political order for that place. This brief review of global farmland preservation policy suggests that Montesquieu's theories still hold sway. England's pastoral heritage is reflected in planning policies that attempt to balance farmland preservation with its growing urban centers. The Netherlands ethos of liberal democracy is evident in its democratized emphasis on agricultural collectives, governed by farmers and rural landowners, coupled with strong planning measures reflecting its monarchical past, while China's authoritarian *esprit* is obvious in its strict classification regimes and heavy-handed oversight.

But as the world's population continues to grow, stretching the limits of its agricultural capacity, and new awareness emerges about the environmental benefits created by farmlands and working forests, the issue of farmland preservation is quickly becoming a global one. Given these global realities, the issue now becomes not how farmland preservation and farmland preservation policy is conducted in individual countries around the world, but how we come together as a collective society to make sure that it is done at a scale that protects us all, no matter the process and policy. Even Montesquieu would agree that the concept of place is a scaled one. And if the challenges of this emerging global society are often wicked ones, presenting no one solution, agrarian policymakers can benefit from learning about the farmland preservation measures of other countries and other places, and adapting them to their own.

Notes

1. "Agricultural Land Protection: An Essential Tool for Fighting Climate Change," American Farmland Trust, 2024, https://farmlandinfo.org/wp-content/uploads/sites/2/2024/04/AFT-ag-land-protection-fight-climate-change.pdf.
2. "Regenerative Organic Agriculture and Climate Change," Rodale Institute, accessed September 4, 2025, https://rodaleinstitute.org/wp-content/uploads/rodale-white-paper.pdf.
3. "Database of Practices in Food System Transformation," United Nations Food Systems Coordination Hub, accessed September 4, 2025, https://www.unfoodsystemshub.org/hub-solution/database-of-practices/en.
4. "Demography of Europe: 2024 Edition," Eurostat, accessed September 5, 2025, https://ec.europa.eu/eurostat/web/interactive-publications/demography-2024.
5. Rachelle Alterman, "The Challenge of Farmland Preservation: Lessons from a Six-Nation Comparison," *Journal of the American Planning Association* 63, no. 2 (1997): 227.
6. "Developing Farmland: Regulations on Land Use," Gov.UK, September 7, 2012, https://www.gov.uk/guidance/developing farmland-regulations-on-land-use#:~:text=the%20new%20buildings.-,Agricultural%20Land%20Classification%20and%20soil%20types,3a%20contains%20a%20similar%20amount.
7. "25 Year Environment Plan," Department for Environment, Food & Rural Affairs, 2018, 9, https://www.gov.uk/government/publications/25-year-environment-plan.
8. "Guide to Assessing Development Proposals on Agricultural Land," Natural England, February 5, 2021, https://www.gov.uk/government/publications/agricultural-land-assess-proposals-for-development/guide-to-assessing-development-proposals-on-agricultural-land.
9. "Conservation Covenants: A New Tool to Support Nature Recovery," Gov.UK, December 19, 2024, https://environmentagency.blog.gov.uk/2024/12/19/conservation-covenants-a-new-tool-to-support-nature-recovery/.
10. Peter Bishop, Alona Martinez Perez, Rob Roggema, and Lesley Williams, *Repurposing the Green Belt in the 21st Century* (University College London Press, 2020), 12.
11. Bishop, *Repurposing the Green Belt*, 21, 156.
12. Charlotte Wright and Emily Coady-Stemp, "PM Reiterates 'Grey Belt' Approach to New Homes," BBC, September 19, 2024, https://www.bbc.com/news/articles/c8elw1zpxewo. Also, Joshua Nevett, "Housing and the Green Belt Puts Politics on Tricky Turf," BBC, July 3, 2024, https://www.bbc.com/news/articles/c0w4ppp25r0o.
13. Alterman, "Challenge of Farmland Preservation," 228.
14. "Natural Ecological Network (NED)," Government of the Netherlands, accessed September 9, 2025, https://www.government.nl/topics/nature-and-biodiversity/national-ecological-network-nen.
15. "Environment and Planning Act," Government of the Netherlands, accessed September 9, 2025, https://www.government.nl/topics/environment-and-planning-act.
16. "European Population Density: Country Rankings," TheGlobalEconomy.com, accessed September 9, 2025. https://www.theglobaleconomy.com/rankings/population_density/European-union/.
17. "Farms and Farmland in the European Union," Eurostat, November 2022, https://ec.europa.eu/eurostat/statistics-explained/index.php?title=Farms_and_farmland_in_the_European_Union_-_statistics.

18 Edwin Alblas and Josephine van Zeben, "Collaborative Agri-Environmental Governance in The Netherlands: A Novel Institution Arrangement to Bridge Social-Ecological Dynamics," *Ecology & Society* 28, no. 1 (2023), doi:10.5751/ES-13648-280128.

19 "Policy on Nature and Biodiversity," Government of the Netherlands, accessed September 9, 2025, https://www.government.nl/topics/nature-and-biodiversity/policy-on-nature-and-biodiversity.

20 "Ownership and Governance of Protected Areas," Department of Climate Change, Energy, the Environment, and Water, May 9, 2025, https://www.dcceew.gov.au/environment/land/nrs/about-nrs/ownership#:~:text=Last%20updated:%2009%20May%202025,Number%20of.

21 Sarah Brugler, "Evaluating Successes and Challenges for Effective Governance of Privately Protected Areas in Australia," *Frontiers in Conservation Science* 4 (December 2023).

22 Forrest Hisey, "Provincial Diffusion, National Acceptance: The Transfer of Conservation Easement Policy in Canada," *Facets* 10 (2025): 1–11.

23 Trevor Budge and Andrew Butt, "Farmland Preservation in Australia," in Wayne Caldwell et al., Eds., *Farmland Preservation: Land for Future Generations* (University of Manitoba Press, 2017), 216.

24 Budge and Butt, "Farmland Preservation in Australia," 220.

25 National Farmers Federation, "Farmers Welcome National Food Plan Commitment," March 4, 2025, https://nff.org.au/media-release/farmers-welcome-national-food-plan-commitment/#:~:text=The%20National%20Farmers'%20Federation%20(NFF,the%20strategy%20delivered%20real%20outcomes.

26 Budge and Butt, "Farmland Preservation in Australia," 225.

27 Michael Buxton and David Mercer, "Pasture Not Pavements: Governmental Planning Failure and the Slow Death of Melbourne's Green Belt," *Land* 13, no. 12 (2024), 17.

28 Alterman, "Challenge of Farmland Preservation," 225.

29 Wayne Caldwell, Sara Epp, Xiaoyuan Wan, Rachel Singer, Emma Drake, and Emily C. Sousa, "Farmland Preservation and Urban Expansion: A Case Study of Southern Ontario, Canada," *Frontiers in Sustainable Food Systems* 6 (2022).

30 David J. Connell, "The Quality of Farmland Protection in Canada: An Evaluation of the Strength of Provincial Legislative Frameworks," *Canadian Planning and Policy* (2021).

31 Nan Wang, Jinmin Hao, Lei Zhang, Wenkai Duan, Yunyang Shi, Jinyi Zhang, and Paruke Wusimanjiang, "Basic Farmland Protection System in China: Changes, Conflicts, and Prospects," *Agronomy* 13, (2023).

32 Tongwei Qiu, Biliang Luo, Shangpu Li, and Qinying He, "Does the Basic Farmland Preservation Hinder Land Transfers in Rural China," *China Agricultural Economic Review* 12, no. 1 (2020): 39–56.

33 Wang et al., "Basic Farmland Protection System in China."

34 Taiyang Zhong, Wei Zheng, Yuntai Zhao, Yu Zhang, and Xianjin Huang, "Farmland Preservation Policies in China and Their Impact on Urban Expansion: A Multilevel Analysis," *Journal of Environmental Planning and Management* 64, no. 14 (2021): 2252–2572.

35 "Countries by Population Density 2025," World Population Review, accessed September 16, 2025, https://worldpopulationreview.com/country-rankings/countries-by-density.

36 "Israel's Birth Rate Remains Highest in OECD, at 2.9 Children per Woman," *The Times of Israel*, September 16, 2025, https://www.timesofisrael.com/israels-birth-rate-remains-highest-in-oecd-by-far-at-2-9-children-per-woman/#:~:text=Israel's%20total%20fertility%20rate%20sits%20at%202.9,OECD%20average%20of%201.5%2C%20the%20report%20said.

37 *Statistical Yearbook of Jerusalem*, Jerusalem Institute for Policy Research, 2024, accessed September 16, 2025, https://jerusaleminstitute.org.il/en/yearbook/. Also, "About the Israel Land Authority," Israel Land Authority, State of Israel Government Portal, accessed September 16, 2025, https://www.gov.il/en/departments/israel-land-authority.
38 Alterman, "Challenge of Farmland Preservation," 234.
39 Ravit Hananel, "The Land Narrative: Rethinking Israel's National Land Policy," *Land Use Policy* 45 (2015), 129.
40 Hananel, "Land Narrative."
41 "Farmland Preservation and Land Use Report, 2021–2023," Ministry of Agriculture and Rural Development (Israel), https://moag.gov.il.
42 "Agriculture in Israel: Statistics and Facts," Statista, December 18, 2024, https://www.statista.com/topics/12937/agriculture-in-israel/#topicOverview.
43 Gabriel Schwake, "The Community Settlement: A Neo-Rural Territorial Tool," *Planning Perspectives* 36, no. 2 (2021): 237–257.
44 Hadas Shadar, "Preservation of Rural Characteristics in Urbanized Villages," *Land* 14 (2025): 1024–1041.
45 Gilberto Camara et al., "Impact of Land Tenure on Deforestation Control and Forest Restoration in Brazilian Amazonia," *Environmental Research Letters* 18, no. 6 (2023): 065005.
46 Claudia Horn, "The International and Local Politics of the Rural Environmental Registry: Brazil's Green Currency," *Development and Change* 55, no. 6 (2024): 1230–1258.

Bibliography

Alblas, Edwin, and Josephine van Zebenk. 2023. "Collaborative Agri-Environmental Governance in The Netherlands: A Novel Institution Arrangement to Bridge Social-Ecological Dynamics," *Ecology & Society* 28, no. 1. doi:10.5751/ES-13648-280128.

Alterman, Rachelle. 1997. "The Challenge of Farmland Preservation: Lessons from a Six-Nation Comparison," *Journal of the American Planning Association* 63, no. 2: 220–243. doi:10.1080/01944369708975916.

American Farmland Trust. 2024. "Agricultural Land Protection: An Essential Tool for Fighting Climate Change." https://farmlandinfo.org/wp-content/uploads/sites/2/2024/04/AFT-ag-land-protection-fight-climate-change.pdf.

Bishop, Peter, Alona Martinez Perez, Rob Roggema, and Lesley Williams. 2020. *Repurposing the Green Belt in the 21st Century*. University College London Press.

Brugler, Sarah. 2023. "Evaluating Successes and Challenges for Effective Governance of Privately Protected Areas in Australia," *Frontiers in Conservation Science* 4. doi:10.3389/fcosc.2023.1291745.

Budge, Trevor, and Andrew Butt. 2017. "Farmland Preservation in Australia." In *Farmland Preservation: Land for Future Generations*, edited by Wayne Caldwell, Stew Hilts, and Bronwynne Wilton. University of Manitoba Press.

Buxton, Michael, and David Mercer. 2024. "Pasture Not Pavements: Governmental Planning Failure and the Slow Death of Melbourne's Green Belt," *Land* 13, no. 12. doi:10.3390/land13121984.

Caldwell, Wayne, Sara Epp, Xiaoyuan Wan, Rachel Singer, Emma Drake, and Emily C. Sousa. 2022. "Farmland Preservation and Urban Expansion: A Case Study of Southern Ontario, Canada," *Frontiers in Sustainable Food Systems* 6. doi:10.3389/fsufs.2022.777816.

Camara, Gilberto Camara, 2023. "Impact of Land Tenure on Deforestation Control and Forest Restoration in Brazilian Amazonia," *Environmental Research Letters* 18, no. 6. doi:10.1088/1748-9326/acd20a.

Connell, David J. 2021. "The Quality of Farmland Protection in Canada: An Evaluation of the Strength of Provincial Legislative Frameworks," *Canadian Planning and Policy* 1: 109–130. doi:10.24908/cpp-apc.v2021i01.14229.

Hananel, Ravit. 2015. "The Land Narrative: Rethinking Israel's National Land Policy," *Land Use Policy* 45. doi:10.1016/j.landusepol.2015.01.015.

Hissey, Forrest. 2025. "Provincial Diffusion, National Acceptance: The Transfer of Conservation Easement Policy in Canada," *Facets* 10. doi:10.1139/facets-2024-0016.

Horn, Claudia. 2024. "The International and Local Politics of the Rural Environmental Registry: Brazil's Green Currency," *Development and Change* 55, no. 6. doi:10.1111/dech.12863.

Jerusalem Institute for Policy Research. 2024. *Statistical Yearbook of Jerusalem*. https://jerusaleminstitute.org.il/en/yearbook/.

Natural England. Updated February 5, 2021. "Guide to Assessing Development Proposals on Agricultural Land." https://www.gov.uk/government/publications/agricultural-land-assess-proposals-for-development/guide-to-assessing-development-proposals-on-agricultural-land.

The Netherlands, Government of. n.d. "Policy on Nature and Biodiversity." https://www.government.nl/topics/nature-and-biodiversity/policy-on-nature-and-biodiversity.

Nevett, Joshua. July 3, 2024. "Housing and the Green Belt Puts Politics on Tricky Turf," British Broadcasting Corporation. https://www.bbc.com/news/articles/c0w4ppp25r0o.

Qiu, Tongwei, Biliang Luo, Shangpu Li, and Qinying He. 2020. "Does the Basic Farmland Preservation Hinder Land Transfers in Rural China," *China Agricultural Economic Review* 12, no. 1: 39–56. doi:10.1108/caer-10-2018-0212.

Rodale Institute. n.d. "Regenerative Organic Agriculture and Climate Change." https://rodaleinstitute.org/wp-content/uploads/rodale-white-paper.pdf.

Schwake, Gabriel Schwake. 2021. "The Community Settlement: A Neo-Rural Territorial Tool," *Planning Perspectives* 36, no. 2. doi:10.1080/02665433.2020.1728569.

Shadar, Hadas. 2025. "Preservation of Rural Characteristics in Urbanized Villages," *Land* 14. doi:10.3390/land14051024.

United Kingdom Department for Environment, Food & Rural Affairs. 2018. "25 Year Environment Plan." https://www.gov.uk/government/publications/25-year-environment-plan.

United Nations Food Systems Coordination Hub. n.d. "Database of Practices in Food System Transformation." https://www.unfoodsystemshub.org/hub-solution/database-of-practices/en.

Wang, Nan, Jinmin Hao, Lei Zhang, Wenkai Duan, Yunyang Shi, Jinyi Zhang, and Paruke Wusimanjiang. 2023. "Basic Farmland Protection System in China: Changes, Conflicts, and Prospects," *Agronomy* 13. doi:10.3390/agronomy13030651.

Wright, Charlotte, and Emily Coady-Stemp. September 19, 2024. "PM Reiterates 'Grey Belt' Approach to New Homes," BBC. https://www.bbc.com/news/articles/c8elw1zpxewo.

Zhong, Taiyang, Wei Zheng, Yuntai Zhao, Yu Zhang, and Xianjin Huang. 2021. "Farmland Preservation Policies in China and Their Impact on Urban Expansion: A Multilevel Analysis," *Journal of Environmental Planning and Management* 64, no. 14. doi:10.1080/09640568.2021.1876000.

18 Farmland Preservation in America

Looking Back, Looking Forward

I started this examination of farmland preservation policy and practice in America with the observation that much has changed over the course of my twenty-plus-year career in the American conservation movement. The use of conservation easements continues to proliferate, but new is the proliferation of government-funded conservation easement acquisition programs, often tied to broader greenbelt and landscape-scale land-use planning initiatives, along with the growth of programs that attempt to use conservation easements not as an end for farmland preservation, but as a means for enhancing access to land and business opportunities for a new generation of farmers.

True, some places like Maryland, California, and New Jersey have been using these programs well before my career horizon, but it is fair to say that "Buy-Sell-Protect" models like the Conservation Fund's "Farm Fund" initiative are new iterations of this model, a shining example of the American conservation movement's spirit of adaptation and innovation. And in a conservative state like North Carolina, the proliferation of county-level farmland preservation positions like mine is truly something new. When I first started in this work, North Carolina counties were happy to leave this conservation easement work to the nonprofit land trusts, and many were quite suspicious about these types of permanent land restrictions. Now it is much more of a collaboration between the two, at least in the rapidly urbanizing portions of the state.

This book was written between the end of 2024 and the beginning of 2026, and what was true for the previous twenty-year period has proven true again over the last year and a half. The beginning of the second Donald Trump administration has ushered in seismic change to United States agriculture and natural resource conservation policy, deeply affecting many of the organizations and farmland preservation practitioners interviewed for this book. If a major theme of the American farmland preservation movement in the twenty-first century is the need to help new farmers access the American agricultural system at a more sustainable scale, the latest Trump administration has turned its agricultural focus toward its trade wars, in some ways a return to the agricultural policies of the Reagan and Nixon administrations that favored industrial-scale farming and manipulating commodity markets;

DOI: 10.4324/9781003644545-21

and in others, a turn back to the protectionism of the Harding/Coolidge/Hoover era.

Federal policy may be influential in the American farmland preservation movement, but it is not dominant, and a central hypothesis of this book is that American farmland preservation is—and indeed, must be—centered in the locus and policy sphere of place, although the scale of that place is still variable. Farmland preservation in Maryland is wildly different than farmland preservation in Wyoming, just as farmland preservation here in the booming central part of North Carolina is vastly different from farmland preservation in North Carolina's vast, rural eastern plain.

One of the evolutions I discussed early in this book was that of an individual rights orientation in American land-use policy and property law to one of collective, social rights that acknowledges the detrimental impacts humans are having on our environment and natural resources. Here, I report somewhat disappointedly, the changes have not been as drastic. Although things *are* changing, especially in rapidly urbanizing places like the American Northeast and metropolitan areas on the West Coast, the changes are gradual ones and often opposed by the powerful development and industrial lobbies, and others economically or ideologically opposed to more sustainable American land use, and their allies in Congress and the state legislatures.

A prominent example has occurred in Chatham County, as I was writing this book. Our county planning department spent a few years crafting a comprehensive development ordinance and zoning code revision that would see minimum lot-size requirements increased in many of our rural areas as an effort to preserve larger parcels of undeveloped land. But just as the new ordinances were set to go into effect, our state legislature passed a statewide law prohibiting these types of rezonings, essentially overturning at the state level our local effort to address the extreme development pressure threatening our rural character here in Chatham County. That our Republican-controlled legislature habitually adopts the mantras of individual rights in their efforts to squash these types of "smart-growth" initiatives is no less surprising than their frequent accord with the powerful lobbies of our state's homebuilder and real estate development groups, who have the most to lose if the ordinance revision ultimately goes forward.

Also somewhat disappointing has been my own work with our local agricultural advisory board (AAB). As I described in Chapter 7, these AABs are local, citizen-led advisory boards required by North Carolina general statutes in order to implement state farmland preservation programs, such as our voluntary agricultural district and our farmland preservation planning process. Part of my job responsibility is providing administrative oversight for this board—scheduling and organizing meetings, compiling meeting minutes, liaising between the AAB and other parts of our county government, etc.

In Chapter 7 I suggested that by providing a regular process for citizen engagement and oversight of local farmland preservation initiatives, these AABs were a good example of the citizen engagement Kemmis calls for in the politics of place. I still think they can be. But in practice, our AAB has fallen victim to some of the same roadblocks and setbacks that frequently hinder these types of citizen engagement and oversight efforts. Member turnover, canceled meetings, called meetings that don't achieve a quorum, unresponsive communications with our elected county officials—indeed, many of the erosions to our social fabric about which Putnam laments in *Bowling Alone*—have all delayed or impeded our AAB's progress.

Ultimately, I finish this book with mixed feelings about the politics of place and its potential for farmland preservation. Like Kemmis writing over thirty years ago, I still believe in the potential for citizen engagement, even its need. I believe that government, attuned to the power of social forces like social capital and sense of place, can facilitate this type of citizen-engaged government, even if effective politics of place must also require engagement from outside the government realm, the type of engagement that can only come from committed community leaders and bridging organizations like nonprofit land trusts. But I also know from my own qualitative experience that this type of alchemy is rare, and that the social forces standing in its way remain as powerful as ever.

The news is not all discouraging. Recently I have become involved with another local group of citizens who are trying to establish community benefit agreements (CBAs) with some local real estate developers. CBAs are voluntary contracts between developers and a coalition of community groups promising benefits to the local community in exchange for the community's support of a new real estate development project. Although CBAs are not specifically a farmland preservation tool—for they typically address affordable housing, funding for community services, and provisions for parks and community centers, among other amenities—they do frequently include open space and farmland preservation measures. For instance, one of the ideas being discussed in our group is requiring developers to contribute money to our local farmland preservation fund; another is assistance in establishing some sort of microloan or grant program to help local farmers access equipment and other small scale farming assets.

My involvement in this CBA group has convinced me more thoroughly than ever before about the importance of leadership in the politics of place. In Chapter 13 I wrote about transformational leadership, the brand of leadership that facilitates citizen governance through heightening consciousness of collective interest. Transformational leaders focus on addressing local issues through a shared-goals orientation that achieves bridging connections across networks and hierarchies of social groups.[1]

The leader of our CBA group appears to possess these qualities. He is experienced in organizing and facilitating group processes and gifted in collectivizing our individual interests and goals into a (mostly) shared vision. Whether our effort is successful or not, his vision and leadership inspire and motivate my own participation in the group, create in me a sense of obligation and commitment to our work, and affirm the findings of resilience scholars like Carl Folke, Per Olsson, and Lance Gunderson (among others), who find that this type of leadership is instrumental, indeed a requirement, for successful governance of complex ecosystem and land-use management issues such as farmland preservation.[2]

It was the legacy of resilience scholars like Folke, Olsson, and Gunderson to which I hoped to contribute with the writing and publication of this book. I came under their influence during my doctoral studies and, obviously, my own perspectives on concepts of social capital, stakeholder process, bridging organizations, and transformational leadership are deeply influenced by theirs. It was the interaction of social capital and sense of place, the social process I have dubbed "cosmic capital," that I hoped was a somewhat novel iteration of their theories, something I could add to their theoretical concepts of social-ecological resilience.

Whether that effort was successful or not, you will have to be the judge. Certainly, my work as the farmland preservation coordinator in Chatham County over the course of the last two and half years has done nothing to diminish my belief in the importance of sense of place. Indeed, I will state that sense of place, and the commitment to environmental stewardship of place it engenders, is the common denominator among all the landowners with whom I am currently working to preserve their farmland. And without the social capital that has coalesced around the effort to preserve our rural landscapes in Chatham County, symbolized as "farmland preservation" but in fact a deeper social longing for a revitalization of rural community, I wouldn't have my job, and probably wouldn't have been inspired to write this book.

But when I submitted the original book proposal to my editors, they asked me to focus as well on real-world examples of farmland preservation, and how they contributed to the theoretical concepts I wanted to explore. Now, as I write these final words, and prepare to submit the manuscript for publication, I must admit I could not be more glad that they did. For it was my interactions with these farmland preservation practitioners, through the interviews they graciously conducted with me, the time and insight they graciously bestowed on me, that I was most dramatically convinced about the ongoing importance of American farmland preservation, and the spirit of dedication and optimism that defines its practitioners. One and all, they not only believed in the importance of preserving American farmland, but also the importance of preserving the American farmer and the rural communities in which those farmers still form cornerstones. Their work, enthusiasm, and commitment inspired me through the long, scrupulous process of writing this book, just as they inspire me still.

And so I close by affirming my abiding optimism for farmland preservation in America. Yes, the structural issues facing the American farmland are "wicked," even overwhelming, but the agrarian *esprit* that has long shaped us as Americans and shaped our collective vision of America, and what America can one day be, is still burning bright. The work of preserving American farmland still has a long way to go; together we will get there.

Notes

1 Nicole Adroin, Rachelle K. Gould, Elin Kelsey, and Priya Fielding-Singh, "Collaborative and Transformational Leadership in the Environmental Realm," *Journal of Environmental Policy and Planning* 17, no. 3 (2014): 360–380.
2 Per Olsson et al., "Shooting the Rapids: Navigating Transitions to Adaptive Governance of Social-Ecological Systems," *Ecology & Society* 11, no. 1 (2006).

Bibliography

Adroin, Nicole, Rachelle K. Gould, Elin Kelsey, and Priya Fielding-Singh. 2014. "Collaborative and Transformational Leadership in the Environmental Realm," *Journal of Environmental Policy and Planning* 17, no. 3. doi:10.1080/1523908x.2014.954075.

Olsson, Per, L.H. Gunderson, S.R. Carpenter, P. Ryan, L. Lebel, C. Folke, and C.S. Holling. 2006. "Shooting the Rapids: Navigating Transitions to Adaptive Governance of Social-Ecological Systems," *Ecology & Society* 11, no. 1. doi:10.5751/es-01595-110118.

Index